The Scented Wild Flowers
of Britain

ROY GENDERS

The Scented Wild Flowers
of Britain

with 32 colour plates
and 42 pen and ink drawings by
MARJORIE BLAMEY

COLLINS
ST JAMES'S PLACE, LONDON

TO THE MEMORY

OF MY PARENTS

W. H. (HARRY) and MAY GENDERS

R 71 17462

582.13

ISBN 0 00 211796 7

First published 1971

© Roy Genders 1971

Printed in Great Britain
Collins Clear-Type Press
London and Glasgow

Contents

Colour Plates

Pen and Ink Drawings

Introduction

'Of all things', wrote Richard Jefferies, 'there is none so sweet as sweet air, and . . . the sweetest of all things is wild flower air'. Samuel Taylor Coleridge expressed similar thoughts in his poem 'To Nature':

> So will I build my altar in the fields,
> And the blue sky my fretted dome shall be,
> And the sweet fragrance that the wild flower yields
> Shall be the incense I will yield to Thee,
> Thee only God.

The wild flowers of the British Isles, which in many cases have a common ancestry with those of northern Europe and North America, are among the loveliest of all flowering plants and possess such extreme hardiness that they are able to survive the most adverse climatic conditions. They are mostly humble little plants, hiding their flowers behind tufts of winter's withered grass and leaves, enjoying the cool moist conditions and often showing themselves only after hours of diligent searching. We love them for their fragrance most of all; the Primrose (p. 139) for its cool sweet mossy scent – the perfume of moist woodlands – and because it is one of the first wild flowers to show itself after the frost and snow of winter have disappeared. The joy of finding the first Primrose is beautifully expressed by the Northamptonshire ploughman-poet, John Clare, in lines written from his cottage at Helpston in 1816:

> Welcome, pale primrose! starting up between
> Dead matted leaves of ash and oak, that strew
> The every lawn, the wood, the spinney through
> 'Mid creeping moss and ivy's darker green;
> How much thy presence beautifies the ground:
> How sweet thy modest, unaffected pride
> Glows on the sunny bank, and wood's warm side.
> And where thy fairy flowers in groups are found,
> The school-boy roams enchantedly along,
> Plucking the fairest with a rude delight:
> While the meek shepherd stops his simple song,
> To gaze a moment on the pleasing sight,
> O'erjoy'd to see the flowers that truly bring,
> The welcome news of sweet returning spring.

Soon follows the Lily of the Valley (p. 205) in its heavily shaded home. John Lawrence, in his book *The Flower Garden* (1726), wrote, 'The Conval-lily is esteemed to have of all others, the sweetest and most agreeable perfume; not offensive or overbearing, even to those who are made uneasy with the perfumes of other sweet-scented flowers'. With its waxy-white bells backed by broad leaves of darkest green, there is no lovelier sight in the half-light of the woodlands.

> The spotless lily, by whose fine leaves
> Be voted the chaste thoughts of virginity;

wrote William Browne in *Britannia's Pastorals*; and Wordsworth sang its praises in his lovely lines to

> The lily of the vale,
> That loves the ground, and from the sun with-holds
> Her pensive beauty; from the breeze her sweets.

Like those other plants that bloom with it, the Lily of the Valley loves best the coolness of a northerly situation, where the early summer sunshine will not dry out its roots nor distil the perfume from its flowers, which it retains in concentrated form for night-flying Lepidoptera.

Sometimes to be found near it is the Scented Solomon's Seal, *Polygonatum odoratum* (p. 208), which comes into bloom before the end of June. It is distinguished from the Common Solomon's Seal by its angular stem and the heavy scent of its dangling bells. Like the Lily of the Valley, it enjoys the coolness of deciduous woodlands. So, too, does the Wild or Woodland Tulip, *Tulipa sylvestris* (p. 211), with its handsome flowers of yellow and green, elegantly pointed petals, and sweet scent which attracts moths for fertilisation: the survival of plant and insect are inter-related, the one cannot survive without the other. If there were no insects there would be no flowers and the flowers with nectar secreted deep down in long tubes are usually those which have the most pronounced perfume, to attract the Lepidoptera – for they alone are able to reach the nectar. The flowers mostly frequent the valleys and woodlands where the warmth brings out the fullness of their scent, and where the butterflies and moths are not troubled by strong winds as in more exposed situations.

In the open meadow the 'freckled' Cowslip blooms, each tiny flower marked with a spot of ruby red which Shakespeare so delightfully described in *A Midsummer Night's Dream*. It is surely one of the loveliest of all our native flowers with a sweet refreshing scent, like that of a baby's breath, but so difficult to describe. With it blooms the Meadowsweet, 'that Queen Elizabeth of famous memory did more desire than any other sweet herb to strew her chambers'. But it was the aromatic scent of its leaves and not the flowers that the Queen enjoyed.

Shakespeare's Musk Rose and the Eglantine (p. 97) flower in the hedgerows where the soft fragrance of springtime gives way to the heavier perfume of midsummer, when the Honeysuckle (p. 184), Milton's 'well-attired woodbine', blooms with its seductive eastern perfume. But whereas *Lonicera caprifolium*, the wild Honeysuckle of southern Europe has a tube some 30 mm in length and is visited only by night-flying Lepidoptera, *L. periclymenum*, native to the British Isles and northern Europe, has a tube only 20 mm long which allows it to be fertilised by bees as well, which visit the flowers during adverse weather when few night-flying moths are about. Of this plant, John Parkinson (botanist to Charles I), writing in the *Paradisus* (1629), said 'although it be very sweete' he did not have it in his own garden, but let it remain in the hedgerows 'to serve the senses of those that travel by or have no garden'.

On the banks of streams and brooks, the Water Mint (p. 157) is aromatic and refreshing. 'The smell of it', wrote Gerard (1597), 'rejoiceth the heart of man for which purpose they strew it in chambers where feasts and banquets are made'.

John Gerard, who published his *Herbal* in 1597, was an apothecary of Nantwich, Cheshire. He was in London soon after Elizabeth's accession and we learn from him that from 1576 he had been in charge of Lord Burleigh's garden in the Strand and also of the Treasurer's garden at Theobald's in Hertfordshire. Gerard's own garden was in Holborn, near Ely Place, and he would therefore be familiar with the famous garden adjoining the London home of the Bishops of Ely, as would Shakespeare who lodged in nearby Silver Street from 1597 to 1604. It is more than likely that Shakespeare knew the apothecary's famous garden well and the plants that grew in it – 'the Crown Imperial and fair Flower-de-luce'.

In 1596 Gerard published a catalogue of the plants which grew in his own garden, the first *Catalogue of Plants* growing in any English garden, and in the *Herbal* which followed he mentions finding all manner of wild flowers around London: Lily of the Valley at Hampstead; Sweet Woodruff in Gray's Inn Lane; Sweet Rocket in nearby Thames-side meadows. In his *Herbal*, Gerard wrote 'If odours may work satisfaction, they are so sovereign in plants and so comforting that no confection of the apothecary's can equal their excellent virtue'.

Through spring and summer the delicious perfumes persist. In the words of William Blake:

> Thou perceivest the flowers put forth their precious odours,
> And none can tell how from so small a centre come such sweets
> Forgetting that within that centre eternity expands
> Its ever-during doors . . .

On dry hills blooms the Wild Thyme, and Marjoram forms spreading mats (p. 160). The matted Pink may bloom here too (p. 63), though it is at its happiest between the stones of an old wall, as on the ruins of Fountains Abbey in Yorkshire or on the keep of Rochester Castle, in company with the Wallflower, where

> . . . flower and herb soon fill the air with innumerable dance,
> Yet all in order, sweet and lovely.

In August, the White or Night-flowering Campion comes into flower (p. 62). It is of the same family Caryophyllaceae, as the Pink and is most powerfully scented in the evening, for it is fertilised by night-flying Lepidoptera which are the only pollinators able to reach the honey secreted at the end of the long narrow tube. The Red Campion, which opens its blooms by day, has a shorter tube so that it is readily accessible to shorter-tongued insects which are attracted by colour, and does not need perfume for its survival.

There is a smell of musk about the hedgerows during July and August, a heavy scent reminiscent of honey. It is a scent which distinguishes the Musk Orchid, *Herminium monorchis* (p. 226), from all other wild orchids. It is also noticeable in the flowers of the Musk or Field Rose (p. 93), of the Elder (p. 182), and is present in the foliage of the Wild Angelica (p. 123) and the Musk Thistle (p. 202), which hangs its head as if in shame but which releases the satisfying scent of musk (or honey) when warmed by the midsummer sunshine. As summer gives way to autumn, the aromatic smell of dying Bracken and of moorland Heather replaces Rossetti's 'sweet, keen smell' of newly cut grass and of the bean fields with their 'heady' cloying perfume.

In the hedgerows, the rich, aromatic fragrance released by dying Wild Strawberry leaves was described by Bacon as one of the most satisfying of all scents at this time of year. Edward Thomas beautifully expressed his delight at the smells of the countryside in autumn:

> Today I think,
> Only with scents – scents dead leaves yield,
> And bracken and wild carrot's seed,
> And the square mustard field . . .
>
> And smoke's smell too,
> Flowing from where a bonfire burns
> The dead, the waste, the dangerous,
> Which all to sweetness turns . . .

In autumn, too, there is the scent of apples, of wild Crabs (p. 103) as they litter the ground, or of those from the orchard, gathered and laid on hay to ripen, and giving a wholesome scent which Chaucer alluded to in *The Miller's Tale*:

> Her mouth was sweet as bracket or the meeth
> Or hoard of apples laid on hay or heath.

The perfume of the Apple is wrapped in history and has long been appreciated as comfort to a tired mind. It was John Key (physician to Queen Mary Tudor) who wrote in 1552 imploring his fellow being 'to smell an old apple for there is nothing more comforting to the spirits'. Writing a century later, Ralph Austen in his treatise *Of Fruit Trees* said, 'sweet perfumes work immediately upon the spirits for their refreshing, sweet and healthful ayres are special preservatives to health'. Chaucer in *The Franklin's Tale* expressed similar feelings when he wrote:

> The odour of flowers and the freshe sighte,
> Wolde have made any heart lighte
> That ever was born.

The fragrance of wild flowers is there in winter too, for on the coldest days of January the Winter Heliotrope will be in bloom, its flowers of palest lilac having the almond-like perfume of cherry pie. It is to be found on roadsides and waste ground, flowering during the bleakest of winter months, often from a leafless plant. It is closely related to Elecampane, *Inula helenium* (p. 190), whose roots when newly gathered smell of ripe bananas, and to *I. conyza* (p. 191), a downy perennial whose roots smell like spikenard, so that countryfolk have given it the name *Ploughman's Spikenard*. It is a rich aromatic smell, referred to by John Clare on more than one occasion:

> Thy horehound tufts, I know them well
> And ploughman's spikenard's spicy smell.

Until recent times, the roots were dried and burnt on ale-house fires, for the scent would counteract the often unpleasant atmosphere. 'Have you smelled the nard on the fire?' asked Ben Jonson. The plant is usually to be found in woodlands, usually of a chalky subsoil, and with it, though mostly in the southern half of England, may occasionally be found the two native Daphnes (p. 106). Both bloom during February and March, *D. mezereum*, bearing clusters of pinkish-purple flowers, and *D. laureola*, flowers of greenish-white, powerfully fragrant. 'The Mezereon Tree' was one of Francis Bacon's flowers 'that do best perfume the air'; of it he wrote, 'the breath of flowers is far sweeter in the air when it comes and goes like the warbling of music . . .' *D. mezereum* with its pink flowers is mostly visited by butterflies, while *D. laureola* is pollinated by moths.

In bloom at the same time is *Mahonia aquifolium* (p. 41), which may not be native to the British Isles but which has for so long been naturalised in thickets and woodlands as to be counted among our natural flora. An evergreen with handsome holly-like leaves, it bears clusters of bright yellow flowers which have the fragrance of Lily of the Valley. Then in March, as Bacon said, follows *Crocus vernus* (p. 221), with its delicately scented flowers, and once again the Primrose blooms, heralding another year of scented flowers.

But not all flowers and leaves have a pleasing smell; some are most obnoxious. To some people Hawthorn blossom is anything but the pleasant flower loved by Robert Burns. The late Walter de la Mare wrote, 'The hawthorn hath a deathly smell' and he was scientifically accurate in his assessment, for the flowers contain the substance trimethylamine, one of the first products of putrefaction, which is also present in herring brine. Hence the somewhat fishy smell of May blossom when inhaled near-to; but this is lightened by the presence of anisic aldehyde, which is also present in the Cowslip and has a sweet refreshing quality. The flowers are fertilised by dung flies like those of other members of the family Rosaceae, the Sorbus, Cotoneaster and wild Pear, which are among the most foul-smelling of all flowers. As if their evil smell was not sufficient to attract the dung flies, nature has provided them with brown and purple anthers resembling decayed meat as an additional means of attraction.

PRACTICAL USES

Of all sweet-scented plants, those most in demand were used for strewing. In the *Polyolbion*, Michael Drayton wrote:

> In the strewing of these herbs . . . with bounteous hands and free,
> The healthful Balm and Mint from their full laps do fly . . .

Most in demand for strewing was the scented rush, Sweet Flag, *Acorus calamus* (p. 233), which grew only in Cambridgeshire and East Anglia. Mentioned in *Exodus*, it is the plant from which the holy anointing oil was made, for its range extends right across northern Europe and Asia. The Saxons knew it as Beewort. Its broad leaves and flower stems release a refreshing cinnamon-like fragrance when trodden upon. It is said that Thomas à Becket ordered his hall at Canterbury to be strewn every day, in spring with fresh May blossom, in summer with sweet-scented rushes . . . 'that such knights as the benches could not contain might sit on the floor without dirtying their clothes'. Though scented rushes formed a cool and fragrant carpet, they had the fault of attracting fleas and so it was usual to burn Fleabane when the rushes were removed and before they were replaced with a fresh supply. Rush Sunday is a custom still maintained at the Church of St Mary Redcliffe, Bristol.

But to the early peoples, scented flowers and aromatic foliage were in great demand for all manner of purposes quite apart from their value for strewing. Corn Mint, *Mentha arvensis* (p. 156), was to be found in every garden, for its ability to prevent milk from turning sour in summer earned it a considerable reputation. Pliny said that 'it will not suffer milk in the stomach to wax sour' so that mint-water was in daily use as a dose for children. Culpeper suggested applying warm rose petals and mint leaves to the head as a remedy for sleeplessness. During Georgian times, finely ground dried leaves of Corn Mint were carried in snuff boxes, and a pinch was taken and inhaled whenever ill-ventilated rooms caused headaches and giddiness.

Pennyroyal, *Mentha pulegium* (p. 156), so named because it was very effective in removing fleas (pulices) from churches and cottages, was also in demand for making stuffings for various meats, hence its country name of *Pudding Grass*. In an old play is written: 'Let the corporal come sweating under a breast of mutton stuffed with pudding'. To countrywomen, a chaplet of Pennyroyal was considered most efficient in refreshing a tired mind on a warm day.

William Coles writing in 1656 said, 'Herbes . . . comfort the wearied braine with fragrant smells which yield a certain kind of nourishment'. Hippocrates classed perfumes with medicines, especially prescribing them for nervous diseases; and Pliny in his *Natural History* mentions, among others, more than sixty remedies to be derived from Mint and Pennyroyal, and seventeen from Violet.

The aromatic foliage of many of our native Labiates was in demand for culinary purposes and to flavour drinks. The leaves of the Ground Ivy, *Glechoma hederacea* (p. 174) were used to flavour and clarify ale as a substitute for hops; and Balm, *Melissa officinalis* (p. 166) was used to make an invigorating drink known as Carmelite Water, which was first made by the nuns of the religious order of that name. For its sweet fragrance it was included among Thomas Tusser's 'twenty-one herbs for strewing', and as Michael Drayton tells us in his *Muses' Elysium* (p. 167), it was used to weave into garlands together with the stems of Pennyroyal. Gerard said of it, 'it driveth away melancholy and sadness'. The fresh leaves, placed in a warm bath, release their refreshing lemon-like fragrance and bring comfort to a tired body.'

The qualities of Wormwood, *Artemisia absinthium* (p. 200), were praised by Tusser on numerous occasions:

> What savour is better, (if physic be true),
> In places infected, than wormwood and rue?
> It is a comfort for heart and the brain,
> And therefore to have it, it is not in vain.

Tusser, a musician, schoolmaster and poet, who as a boy sang in the choir of old St Paul's, published his *Five Hundred Points of Good Husbandry* in 1573 and in it he gives practical advice on the use of many scented flowers and leaves, including the Carnation and Pink, Violet and Primrose, Wallflower and Columbine. Tusser traded in sheep and oxen and also in grain though with little profit to himself. He died in 1580 a poor man, and was buried at St Mildred's-in-the-Poultry.

The fragrant roots of certain plants were also used in the home. From those of *Sedum rosea* (p. 104), a rose-water was obtained which was used for washing the body after bathing, while from the scented roots of the Sweet Flag, *Acorus calamus*, a perfumed talcum powder was made. The roots of Elecampane, *Inula helenium*, were dried and mixed in pot-pourris, valued because their aromatic fruity scent of ripe bananas assumes a violet perfume with age.

Of all the scented flowers used for flavouring food and wine and to counteract the musty air of manor houses and cottages, none were held in higher esteem than the Violet and Pink. Tusser includes the Violet for strewing and under the heading *For those who have no garden,* suggests cultivating them in pots to place in windows, together with fragrant Sweet Williams, Wallflowers, Stocks and Cowslips (p. 136). In England at the time, Violet flowers were cooked with meat and game and were placed in salads to be eaten raw. In his essay *Of Gardens* Francis Bacon said 'that which above all yields the sweetest smell in the air is the violet, especially the double white violet which comes twice a year, about mid-April and again, about Bartholomew-tide'.

To flavour drinks no flower was more popular than the Clove-scented Pink (p. 64), the Gilliflower of Chaucer's day which grew in the gardens of most ale-houses. The poet referred to it in his *Prologue to the Canterbury Tales* (1386). In *The Tale of Sir Topas* he wrote:

> And many a clove gilofre
> And notemuge, to put in ale,
> Whether it be moiste or stale . . .

We can imagine the Canterbury Pilgrims gathered in the yard of the Tabard Inn in Southwark before their departure, with flowers of the Clove-scented Pink floating in their jars of ale and wine and imparting the rich clove perfume to the drinks. For this reason, the flowers were known as *sops-in-wine*, as in Spenser's lines from *The Shepherd's Calendar*:

> Bring hither the pinke and purple cullambine,
> With gilloflowres;
> Bring coronations and sops-in-wine,
> Worn of paramours.

Until the advent of sugar, both Violets and Pinks were in demand for sweetening as an alternative to honey. Later, the blooms came to be dried and crystallised by dipping the flowers in a solution of gum arabic and rose-water, then sprinkling with fine sugar. The stems of *Angelica sylvestris* (p. 123) were candied in a similar way or were treated with honey and stewed with rhubarb, for the musky aroma was much appreciated. The seed of Angelica, too, has a musk-like scent. In his *Calendar for Gardening* (1661), Stevenson suggested that one should 'every morning, perfume the house with angelica seeds, burnt on a fire-pan or chafing dish of coales, than which nothing is better'. Parkinson, writing thirty years earlier, had said that 'the whole plant, leaf, stem, root and seed is of an excellent comfortable scent, savour and taste'. The roots, which are also aromatic, are used in the making of Vermouth and the seeds to flavour Chartreuse.

Fennel (p. 120) was a plant of similar qualities, for all its parts are fragrant. The seed with its scent of aniseed was used to flavour drinks and from it an excellent sauce was made, to serve with fish. It was Falstaff, in the Boar's Head tavern in Eastcheap, who replied in answer to Doll's question as to why old Poins was so fond of Prince Hal's company, 'And he (too) plays at quoits well and eats conger and fennel . . . and swears with a good grace'. Because of its hay-like smell the Romans named it *Foeniculum*, and made it into wreaths to be worn as an emblem of flattery, a custom which persisted in England until Tudor times. Longfellow, in 'The Goblet of Life', well described its many virtues:

> It gave men strength and fearless mood,
> And gladiators fierce and rude
> Mingled it with their daily food:
> And he who battled and subdued
> A wreath of fennel wore.

FLOWERS AND THEIR POLLINATORS

Nature has decreed that the survival of both plant and pollinator be inter-related. The flower, with its nectar, colour and scent, can exist only in relation to the visits it receives from insects for its pollination, and in turn each plant provides the means for the survival of its pollinators as well as its own. The first plants to inhabit the earth, those bearing cones and catkins, relied on the wind's agency (and to a lesser extent on water) for their pollination, as there were few insects to do the work. Darwin described how he found the ground near St Louis, Missouri, thickly covered with pollen grains, borne by the wind from pine forests some 400 miles away. Conifer pollen is light and dusty and is made even more readily airborne by the swellings of the cuticle of the outer covering. Just before pollination is due, the female cones elongate to admit the pollen grains which are held in place by a sticky fluid. Later, the seeds are released, but only during dry weather, for the cones are sensitive both to wet and dry conditions and open only when the weather is suitable.

Among the earliest forms of plant life, present in Tertiary formations, were the catkin bearers, those of the Orders Salicales, Juglandales and Fagales, represented by the Poplars, Willows, Walnuts and Birches, familiar trees of the British countryside. These rely upon wind to disperse the pollen grains of their scentless pendulous catkins, though several species of Willow with their more erect catkins rely upon insects of all types for their pollination. Of the catkin-bearing families, several have resinous foliage, resembling the conifers. This was a primitive additional feature to provide protection from browsing animals; and from the resinous exudations of the bark, early man obtained health-giving substances for his own use. From the bark of the Birch a nutritious beer is obtained and from the Balsam Poplar, a substitute for friar's balsam, while the bark of the Osier, *Salix purpurea*, yields the medicinal glucoside, salicin.

Development of the insect-pollinated flower was evolved by degrees as additional species of insect came to inhabit the earth. In the family Loranthaceae, which includes the Mistletoe and other semi-parasitic shrubs and which includes more than 1000 species and 27 genera scattered throughout the world, pollination is by wind and by birds which seek the nectar secreted at the base of the perianth tube. The Mistletoe, *Viscum album*, is visited by the more primitive forms of Diptera (flies) and it is of interest that the Magnolia and the closely related Drimys of the Order Ranales, which with their resinous wood and foliage show similar primitive characteristics to the conifers, are pollinated by Coleoptera (beetles), one of the earliest insects to inhabit the earth.

Closely related to Ranales in the hypogynous flowers with free sepals, petals and stamens is the Order Papaveraceae which includes the Red Corn Poppy, a familiar flower of the British countryside. Though adapted for self-fertilisation by the anthers shedding their pollen on the radiating stigmas, the flowers, like those of the Magnolia, are also visited by beetles searching for pollen. Flowers of the near-related Nymphaeaceae family, which includes the White and Yellow Water-lilies, are also pollinated by beetles of the genus *Meligethes*, which creep about the flowers in search of honey, possibly attracted (as in Magnolia) by the fruity perfume of the flowers.

For the same reason that they are pollinated by wind, grasses are also scentless, though the chemical substance coumarin is to be detected in newly mown meadow grass as it begins to dry, a sweet familiar scent which is present also in Woodruff, in the leaves of the Man Orchid, and in the flowers of Sweet Alison (p. 44). Those flowers which are mostly self-pollinating are also devoid of any scent, for they do not find it necessary for their survival. Again, they were among the earliest plants to inhabit the earth.

The inconspicuous flowers of the Order Polygonales, which includes the Dock, are wind-pollinated but are also adapted to pollination by the most primitive insects. The flowers of the family Scrophulariaceae are also visited by insects, but these are not necessary for their fertilisation, for Darwin discovered that when he covered flowers of *Verbascum thapsus* (p. 151) to exclude all insects they set as many seeds as flowers which were left uncovered. In the absence of insects, self-fertilisation is possible, as both anthers and stigma ripen together, with the stigma in line with the fall of pollen.

Although humble-bees visit the flowers of *Digitalis* (Foxglove), self-pollination can and often does take place. The anthers retain pollen until the stigmatic lobes separate, then as the corolla falls away, the pollen comes into contact with the stigma. In *Hypericum* (p. 58), which has resinous foliage, it may be observed that when the flower is open the styles pass outwards without their coming into contact with the anthers, but contact is made immediately the flower begins to fade and close up, when self-fertilisation occurs. In *Malva rotundifolia*, which is rarely visited by insects, self-fertilisation takes place, for the anthers remain extended and are touched by the sides of the curling stigmas. In none of these flowers is there any degree of perfume at all.

Certain flowers are capable of changing their scent both during the day and at night. An example is the native Pyramid Orchid, *Anacamptis pyramidalis* (p. 232), which emits a clove-like carnation scent by day and is visited by butterflies, while at night it emits an unpleasant fox-like smell, possibly to prevent its visitation by night-flying moths. In *Gymnadenia conopsea*, however, the tube is so narrow that the honey is accessible only to certain night-flying Lepidoptera and so it releases its sweet perfume only after dusk.

The scent of flowers usually takes on a different and less pleasant quality after fertilisation, as soon as the flower begins to fade. The smell of damp fur emitted by the flowers of *Orchis mascula* develops from a pleasant vanilla scent after fertilisation, and it would appear that the transformation acts as a protective device, warning the pollinators against visiting the flowers which have now lost all their attraction.

The flowers which appear solitary are usually most fragrant, for those in umbels or heads, e.g. Cruciferae, Umbelliferae, Compositae, attract by appearance and have little or no need of scent. Also, pollinating insects are able to pass more quickly from one flower to another and so few are missed. This would not be so for solitary flowers which are often at a distance from each other and must attract either by brilliance of colour or by their powerful scent.

Flowers which open at night and are fertilised by night-flying Lepidoptera are assisted in attracting the attention of moths by their delicate appearance, being white or pale yellow, and so great is the economy of nature that by remaining closed by day, they retain the full strength of their scent for release in the cool of the evening. This scent is available only to moths, for no other insect visits flowers at night.

Flowers fertilised by Lepidoptera (butterflies and moths) are the most attractively scented, relying on perfume rather than colour, though butterflies which are active by day do seem to be attracted to flowers of pink or mauve-pink colouring, e.g. *Daphne mezereum, Dianthus plumarius*, and *Buddleia davidii* (pp. 106, 63, 145), which they occupy for hours at a time.

Most advanced in their adaptation for visiting flowers and obtaining their honey are the Lepidoptera, while the flowers they visit have a quality far exceeding all others. The butterfly (and to a lesser extent the moth) is known as the *flower of the air* not for its beauty but for its fragrance. This characteristic was first observed by Fritz Müller who found that scent varies in different species, some smelling like vanilla, some like Jasmine, others like Honeysuckle and in fact butterflies may visit only those flowers smelling like themselves. Their scent is produced by an essential oil, present in the flower too, which can be extracted with alcohol. It is secreted in tufts of hairs which lie in a groove on the margin of the wings or on the front legs, and its evaporation is prevented by a waxy substance. In some species the scent hairs lie to the front of the abdomen. During mating, the male may release a scented dust over the female or may brush against her with his scented hairs, producing as it were a secondary sexual characteristic. While butterflies may remain on the flowers of their choice for hours, night-flying Lepidoptera will fly from flower to flower in rapid succession, thus fertilising large numbers whenever weather conditions permit; for their proboscis and mouth are arranged to give the minimum amount of delay in entering the flower and removing the honey.

A far larger number of flowers are visited by bees than by any other insect. Bees are attracted to certain flowers by means of distinguishing marks (usually yellow), like the rays of a Pansy or the bright yellow eye of the Forget-me-not, and if it concerns them at all, scent is of secondary consideration in the bees' search for honey. Though bees are well able to distinguish colours they are red colour-blind, unable to distinguish red from green, and so concentrate their attentions mainly on flowers of blue colouring (with white or yellow guide lines) and also those coloured yellow. In the British Isles most blue flowers, which are rare in nature, are scentless. They are mostly members of the families Campanulaceae, Boraginaceae and Labiatae and though the latter would appear to have scented flowers this is not so, for the familiar aromatic scent is emitted only by the green parts of the plant, including the bracts of the flowers.

Several members of the family Ranunculaceae bear blue flowers which are attractive to bees, and Darwin has shown that *Bombus* (the humble-bee) will fly from one blue-flowered Larkspur, *Delphinium consolida*, to another which shows only a faint tinge of blue in the half-opened buds, neglecting other blue flowers on the way. That the flowers are quite devoid of scent clearly shows the bees' keen powers of colour discrimination. That bees work by sight was confirmed by Darwin when, on a warm day, he observed *Apis* (the honey-bee) visiting the blue flowers of *Lobelia erinus* and upon removing the petals noted that the bees did not visit them again.

Possibly the most primitive type or species of Hymenoptera is *Prosopis* which may derive its origin from the sand wasps. These insects are almost hairless and claim admittance to the family only in the manner in which they feed their young. They have a peculiar odour about them and do not visit the same flowers as the humble- and honey-bees, but concentrate their energies on flowers of a dull yellow colour which emit much the same peculiar smell as themselves. They are to be found on several of the Achilleas; on the Scented Mayweed, *Matricaria chamomilla* (p. 194), on *Ruta graveolens* and other plants of unusual smell and where the honey is readily accessible, for the insects have only a short tongue. In the species *Andrana* and *Halictus*, the tongue is longer and is hairy, while the body is also hairy, especially the outer part of the tibiae of the hindlegs – the young are fed with the pollen which adheres to the hair. In *Bombus* and *Apis* this hairy covering is developed further, while the tongue is also extended so that honey may be obtained from those tubular flowers which are out of reach of all other insects except the Lepidoptera (which concentrate on flowers with scented attractions). Bees work fast, moving from one flower to another with the proboscis extended, gathering honey and pollen on their body hairs, and, as Aristotle observed 2000 years ago, to save time they will visit flowers of the same species for as long as possible before moving on to another.

Lower down the scale, the Diptera (flies and midges) with their short tongues are able to gather nectar only from those flowers where it is produced in shallow cups. The exception is the species *Rhingia* which has a longer tongue and is able to discover nectar which is more deeply hidden.

Most flowers visited by Diptera usually have an unpleasant smell. They include many Compositae and some are so unpleasant as to be attractive only to *Lucilia* and *Sarcophaga*, the so-called dung flies which frequent flowers containing trimethylamine, a substance found in the first stages of putrefaction, which is present in flowers of the Hawthorn, Rowan and Cotoneaster.

A species of midge known as *Psychoda phalaenoides* fertilises the flowers of *Arum maculatum*, attracted by the heat it generates and by its evil smell. As the pale green sheath unfolds in spring, the spadix begins to heat, sometimes reaching a temperature 20° above that of the surrounding air. At this stage it releases, when touched, a smell of stale urine which the midges are unable to resist. The smell is present only for a day while pollination takes place, after which the spadix also loses heat.

Among the more primitive insects are the Coleoptera, the beetles, especially those of the genus *Meligethes* which are able, by their small size, to enter the half-opened buds of flowers and lick the honey and devour the pollen where the honey is too deeply hidden. The flowers visited by Coleoptera may be divided into two groups:

(*a*) those which have a fruity perfume, e.g. *Nuphar lutea, Genista hispanica, Magnolia* and *Rosa*; (*b*) flowers which are scentless and which are visited by beetles in search of warmth and dry pollen, e.g. *Digitalis, Aconitum, Papaver.*

From this it would appear that brilliance of colour rather than scent is the chief attraction. The scentless Field Scabious, *Scabiosa arvensis,* conceals its honey in pin-cushion heads composed of numerous flowers with tubes 6–8 mm in length, and the smaller Coleoptera congregate in these and are able to reach the honey at the base of the tubes.

The closely related field bugs, Hemiptera, are also known to frequent flowers, mostly Compositae and Umbelliferae, in search of honey. *Pyrocoris aptera* will enter the florets of the Dandelion, attracted by the unpleasant smell and upon withdrawal is quite noticeably covered with pollen, thus proving itself to be an effective fertiliser.

COMPOSITION OF SCENT IN FLOWERS AND LEAVES

Scent is the oxidisation of the essential oils of flowers and leaves. In flowers the essential oil is found in the epidermic cells of the petals or in the sepals or bracts which may be present as petal replacements. This oil is usually to be found on the upper surface of the petals, but is present on both surfaces while the petals are concealed in the bud. It is not found on the anthers, though they may absorb the perfume in the same way that Auricula leaves absorb the almost intoxicating perfume of the flowers. The most powerfully scented flowers are those with thick velvet-like petals which prevent much of the oil from evaporating until the flower has actually perished. Scented flowers in the double form, e.g. the double forms of the White Rose, are especially fragrant and retain their scent longer than single flowers.

The essential oil is present in the petals in an inert form, stored as a mixture of oil and sugar known as a glucoside and it is not until a state of fermentation has begun to take place that the perfume can be released. This cannot take place while the flower is closed, which explains why the most fragrant flowers are those which open only at night and whose scent is released very slowly in the cool night air.

It is the chlorophyl, the first substance to appear in the evolution of the flower, which produces the essential oil, in inverse ratio to the amount of pigment in the flower. This explains why scarlet and golden-orange-coloured flowers are usually devoid of scent. These belong mostly to the tropics and are pollinated by humming-birds which have little or no sense of smell. It also explains why white flowers (without pigment) are the most fragrant, e.g. Lily of the Valley, the White Violet, Jonquil and *Daphne laureola*, and furthermore why it is that when colour is bred into flowers perfume is usually lost, while retained in the wild or original form. Scent is a substitute for colour and where colour is intensified scent (or smell) is diminished. The double scarlet Hawthorn, for example, is entirely devoid of any unpleasant smell.

The essential oil is secreted in minute droplets which evaporate when directly exposed to the air to give off a sweet- or unpleasant-smelling vapour which is most noticeable on a warm, calm day. Scent is a waste product of the plant, and for this reason is stored in containers away from the living cells from which the protoplasm has disappeared. In the same way, an animal secretes its smell into a cavity formed around the lanoline glands.

Essential oils are insoluble in water and it was not until the end of the fourteenth century that alcohol was discovered to be a suitable solvent. The alcoholic solution is known as an *essence* and the essential oil is called an *attar* from a Persian word meaning smell or scent.

The production of essential oil continues only while the flower remains alive and as soon as a drop of oil has been released by evaporation, another begins to form. These minute drops evaporate when exposed to the air and give off a sweet or unpleasant smell depending upon the chemical composition of the flower.

In leaves, the essential oil is stored in numerous ways. In the leaves of *Hypericum hircinum*, the oil is stored in deeply embedded capsules while in those of Thyme it is nearer the surface – so near in fact that the scent is released by the sun when the foliage is dry. In Myrtle leaves, the essential oil is so deeply embedded in its cells that it is released only with considerable pressure.

With the Balsam Poplar, the oil is secreted on to the buds where it is fixed with a sticky wax-like substance to prevent its rapid evaporation. If the buds are held over a low flame, the fragrance is released.

The essential oil of a flower or leaf is composed of various chemical compounds. Of these, the most common are esters, formed by the combination of an acid and an alcohol. The fruity scent of certain flowers is due to the combination of acetic acid (which gives the unpleasant smell of perspiration to Valerian root and to Elder bark) and ethyl alcohol which produces ethyl acetate. It is this which gives the fruity smell to the Yellow Water-lily, *Nuphar lutea*, and to Magnolia blossom. The alcohol, borneol, is present in leaves as borneol acetate, and also in the leaves of many conifers with its familiar pine scent. It is eucalyptol which gives the rich herby scent to Wormwood and Yarrow and, combined with the bergamot-scented linalyl acetate, gives

Lavender its familiar scent. The characteristic smell of Mint is due to the alcohol, menthol, while coumarin gives the smell of newly dried hay to Woodruff and Sweet Alison.

In flowers, this chemical composition is more complex. Thus in certain lilies benzine compounds are present, usually as benzyl acetate in association with benzyl formate, and with the nitrogen compound methyl anthranilate. Indol is also present. The ester, methyl anthranilate has an orange-like scent and is present in orange blossom, in Lily of the Valley and in most heavily scented flowers. Methyl indol, the active principle of civet (scatol), an excretory compound, present also in the early stages of putrefaction, is found in the 'heavy-scented' flowers as well. It was the poet Cowper who said he was unable to enter into conversation where there was civet present and there are those who find a concentration of lilies (or Jonquil) equally disturbing. In Lily of the Valley, indol is present in diluted form and is tempered with balsamic compounds which are also present in the flowers of the Hyacinth, the Bluebell and White Jasmine. It is less pleasant in the flowers of the Privet.

The spicy clove-like scent of Pinks is due to the presence of methyl eugenol which is also present in cloves. Geraniol, the chief constituent of attar in roses, is present in many leaves, particularly those of Rose-leaf Pelargoniums, so popular as house plants during Victorian times. It is also present in the root of *Sedum rosea*, the Rose-root of cottage gardens, and combined with eucalyptol, in Lemon-scented Thyme and Balm, *Melissa officinalis*.

Aldehydes, which by oxidation become alcohols, provide the aniseed undertone to Cowslips and Hawthorn blossom, being present in the form of anisic aldehyde and in certain leaves as citral, which is closely related to geraniol. The two in fact are usually found together as in the Eastern Lemon Grass, *Cymbopogon*, and in the stems and leaves of *Acorus calamus*. While predominating in the Rose, geraniol is augmented by perhaps as many as ten or more substances which combine to make one of the considerable number of rose perfumes. Indeed, of the five roses native to the British Isles, each bears a bloom of different scent: that of the Dog Rose is fruity whilst *R. arvensis* smells decidedly of musk (hence its name of the Wild Musk Rose which Shakespeare knew well).

The violet scent, to be found in numerous plants is due to a ketone; ionone in the Violet flower, irone in orris root which is the root of the Florentine Iris, *I. florentina*. The same fragrance may also be detected in the flowers of *Acacia dealbata* and *A. farnesiana* which grow semi-wild in S Cornwall and on the French Riviera. It is also present in the double form of the Wallflower, the yellow Harpur Crewe, in *Hesperis matronalis*, in *Matthiola incana*, and in the closely related Mignonette, *Reseda odorata*, a native of Egypt but well known in British gardens and occasionally found growing wild. Differing only slightly botanically and of the same plant order, is the Sweet-scented Violet.

It may be observed that the same scent is present in all those members of a plant family in which there are scented attractions. For example, the clove perfume of the Pink is also present in a slightly diluted state in the Night-flowering Campion, *Lychnis alba* (p. 62) and in its fullness in the Night-scented Catchfly, *Silene noctiflora* (p. 61), each being of the same family Caryophyllaceae. With the legumes, the scent of Laburnum flowers is almost identical to that of the Sweet Pea and the Field Bean (p. 84), a scent of vanilla with undertones of lemon. John Clare had observed this quality:

> Each different scent possest by different tribes,
> Sense easy feels, but ignorance describes.

The amines provide the unpleasant smells of flowers and leaves. They are carbon compounds of ammonia and are present in Hawthorn and in Sorbus blossom as trimethylamine and propylamine; also in the leaves of Dog's Mercury and the Birthwort, *Aristolochia clematitis* (p. 124). The amines have the unmistakable smell of decaying fish and are present in herring brine.

Compounds of sulphur are to be found in concentrated form in the wild Garlics and in less degree in the leaves of Water Cress, Garlic Mustard and other Cruciferae, especially the Stinkweed, *Diplotaxis muralis* (p. 43).

Whereas the primary object of the scent of flowers is to attract insects for their fertilisation, the essential oil of leaves has the opposite effect, its purpose being to prevent bacteria and insects from attacking the plants. In warm countries, the aromatic vapour released from the plant by the heat of the sun acts as a protective covering for the plant, preventing the moisture from evaporating too rapidly from the foliage. It may also protect the plants from browsing animals.

Most of the Labiates common to the British Isles which usually inhabit barren, open places, fully exposed to the sun, have a protective covering of hairs and are pleasantly aromatic. Examples are *Glechoma hederacea*, *Marrubium vulgare* and *Ajuga chamaepitys* (pp. 174, 175, 176).

Many plants which have a hairy covering smell remarkably like musk. One is the Labiate Clary, *Salvia horminoides* (p. 169), whose leaves have a grey appearance due to the silky hairs. Very little of the musk smell is noticeable until the leaves begin to dry, when it becomes most pronounced. However, the fresh leaves yield an essential oil which has the same musk scent.

The Musk Mallow, the Musk Storksbill and the Musk Thistle, all have their foliage covered with short hairs and possess the same smell of musk as the Clary. It is interesting to note that when *Mimulus moschatus,* the Musk plant of cottage gardens, reverted to its original almost hairless form (which happened throughout the world at the same time – in 1913) it also lost the rich musky perfume for which it had long been famous. Of the same family Scrophulariaceae, the Common Mullein, *Verbascum thapsus,* has velvet-like leaves thickly covered with white woolly hairs and when it is handled it releases a rich musky smell.

Most of the Labiates yield an essential oil with powerful antiseptic qualities, for it may be assumed that as these oils are stored by the plant in containers away from the living protoplasm, they must be poisonous to the plant and will also be poisonous to bacteria invading the plant.

The essential oil of Thyme leaves is known to have twelve times the antiseptic power of carbolic acid, and Professor Chamberlain has demonstrated that it will kill typhoid bacillus within fifty minutes. The oils of Lavender and Rosemary are equally powerful in their antiseptic qualities.

The burning of incense in churches was originally performed to combat disease and is continued until this day as a symbolic tradition, in the same way that the boys of Christ's Hospital walk before the members of the Skinner's Company with bouquets of scented flowers and herbs to ward off the plague and other obnoxious diseases.

Most evil-smelling plants are poisonous and the smell they emit may be nature's way of warning mankind and animals against touching them. The Hemlock is poisonous in all its parts and is our only Umbellifer with a really unpleasant smell, while the lurid purple markings on the stems also act as a repellent.

Most members of the Nightshade family, Solanaceae, are poisonous and emit a most unpleasant smell, none more so than the Henbane, *Hyoscyamus niger,* with its clammy leaves and creamy-brown flowers with purple veining, the whole plant having the same evil-looking appearance as the closely related Deadly Nightshade.

CLASSIFICATION OF FLOWER AND LEAF SCENTS

No classification of flower and leaf scents has ever been made, the reason being that there is a vast variation in perfumes. Due to the different pigments present in the olfactory mucous membrane, scents can vary in the effect they have upon the senses. Again, perfumes change with dilution. A Lily inhaled near-to has a very different effect upon the senses than one inhaled from a distance, when the unpleasant indol undertones have become dispersed in the atmosphere. Hawthorn blossom inhaled from a distance is pleasant, the aniseed undertones predominating, but near-to the sweetness disappears to be replaced by the smell of decaying fish (trimethylamine).

It was not until 1893 that the first attempt at classifying flower scents was made. This was done by an Austrian, Count von Marilaun, who arranged them into groups according to the chemical substance that predominated in their essential oils. He classified the scents into six main categories – Indoloid, Aminoid, Benzoloid, Turpenoid, Paraffinoid and Honey. Several, however, were unsatisfactory, as in the Turpenoid group, which comprises lemon-scented flowers, for it was apparent that the scentless turpines contributed nothing to the lemon scent. It was also considered necessary to divide the Benzoloids into three groups with the result that there are now ten main groups, though all are connected to each other by the presence of the various chemical substances in more, or less, concentrated form in several groups.

(1) *Indoloid Group*

Here indol is present in a concentration great enough to produce the smell of decaying flesh. The group is mostly represented by those plants of the Aristolochiaceae and Araceae families, the latter represented in the British Isles by the Cuckoo-pint, *Arum maculatum*. Many other flowers contain indol, though in less concentrated form and they are included in what is known as the Heavy Group, for they are made more pleasant by the presence of other chemical substances. The flowers of Group (1) are usually purple or brown to give the illusion of decaying flesh. Pollination is by Diptera (dung flies and midges).

(2) *Aminoid Group*

The flowers are usually dingy white or cream-coloured with brown or purple anthers, the colour of decayed flesh, and are borne in dense inflorescences, thus attracting the insects by sight as well as by scent. Included here are *Crataegus* (Hawthorn), *Pyracantha*, *Sorbus*, *Cotoneaster*, tree-like plants of the Rose family, as well as Elder, Privet and several Umbelliferae, including the Giant Fennel and Hemlock, which have the fetid smell of decaying fish. The Hawthorn is less unpleasant, for anisic aldehyde is also present to provide a sweet aniseed undertone. The unpleasant smell of decaying fish is due to trimethylamine and propylamine, present in the early stages of putrefaction and also in herring brine. It is an ammonia-like smell and like the Indoloids, attracts Diptera. It may be said that the ammonia-like smell of Privet, *Ligustrum vulgare*, is absent (or almost so) from the closely related Lilac, occasionally found naturalised in the British Isles and included in the Heavy Group. Some people find this the most attractively scented of all flowers.

The unpleasant ammonia-like smell is also found in the leaves of Dog's Mercury and the Stinking Goosefoot, due to the presence of the same compounds.

(3) *Heavy Group*
Those flowers which comprise this group are among the most power-fully scented of all. They are mostly of sub-tropical regions – of the valleys of Central China, of South America, and also of the Middle East – and include such plants as the Madonna Lily, Jonquil, *Syringa*, *Philadelphus*, White Jasmine and the Tuberose. All are white-flowered, with thick wax-like petals. Methyl indol is the chief substance (also present in the civet as scatol) and is very disturbing if inhaled in con-centrated form. The more pleasing undertones are provided by benzyl acetate, which has the scent of jasmine and methyl anthranilate, which is scented of orange. These substances separately or combined provide a sweetness which in many flowers, e.g. Lily of the Valley, obliterates the presence of indol. Pollination is mostly by moths, for the honey is secreted at the base of long tubes, while the flowers are most powerfully scented at night.

Rather less heavily scented are those plants which inhabit the wood-lands of the British Isles: Lily of the Valley, *Daphne laureola*, *Lonicera periclymenum*, *Linnaea borealis* and *Moneses uniflora* (*Pyrola uniflora*). All bear white or pale cream flowers and attract night-flying Lepidop-tera with their powerful scent. The Butterfly and Scented Orchids, of woodlands and grasslands, may also be included in this group.

(4) *Aromatic Group*
The flowers of this group contain the essential oils found in small amounts in scented leaves, so the undertone is usually light and aromatic rather than heavy when indol is present. The balsamic scent is present in flowers of the Broad Bean, in Night-scented Stock and in the Bluebell, while the clove scent, due to eugenol is present in Pinks and Carnations, in the Catchflies of the same family and in the Vibur-nums; for although their scent is heavy an aromatic clove-like perfume is also present. The same scent also occurs in the Clove-scented Broomrape and in the Fragrant Orchid, *Gymnadenia conopsea*. The roots of *Geum urbanum*, too, are clove-scented.

The scent of aniseed (anisic aldehyde) is present in flowers of the Primrose and Cowslip, while in the roots of *Primula veris* are two glucosides, one yielding an oil smelling slightly of Wintergreen, while the other smells powerfully of aniseed, the same scent that is present in the flowers. It occurs also in the stems and is released when these are cut.

The vanilla scent may be included in this group. It is the most pleasing of all flower scents but is present in only a few flowers of the British Isles including *Clematis vitalba*, and in the Laburnum, Tree Lupin and Field Bean, long naturalised in these islands. Many of the legumes bear vanilla-scented flowers.

The scent of almonds is to be detected in *Convolvulus arvensis*, in the

Flowering Rush and in *Spiranthes autumnalis,* each of which carry the scent of Heliotrope in less concentrated form.

Throughout the group, the flowers are of pink colouring or are white shaded with pink, and they have their honey deeply concealed. Almost all bloom during the hours of daylight and are visited by butterflies rather than moths, with the exception of the Night-scented Catchfly. The leaves of the Meadowsweet may be included here and have a different smell from the flowers.

(5) *Violet-scented Group*

Few flowers diffuse the violet perfume which rapidly causes fatigue of the olfactory nerves so that the scent 'comes and goes' and one needs a rest from the flower to appreciate it afresh. A cool, moss-like quality accompanies the excessive sweetness and this makes the scent most pleasing. It is pronounced in *Iris reticulata* and in *Asparagus tenuifolius* but is present only in the Violet native to the British Isles. The Violet produces its own self-fertilising flowers and does not rely on insects for its pollination. Its scent may in time fade away for, in the same way as humans, pollinating insects may lose it through fatigue.

(6) *Rose Group*

The rose scent is composed chiefly of the essential oil geraniol combined with a fruity substance; thus we have roses smelling of oranges, apricots and raspberries. It is closely allied to the Aromatic Group, though the scent has a lighter quality. Flowers of the Burnet Rose have fruity undertones, while the leaves of Sweetbriar, *R. rubiginosa,* have a lemony smell when handled. Of other native flowers, the Lady's Slipper Orchid has a slight rose scent when newly opened.

As with most flowers of fruity scent, e.g. *Magnolia* and *Nuphar lutea,* roses secrete no nectar and are also visited by beetles. The roots of *Sedum rosea* are rose-scented.

(7) *Lemon Group*

This is closely related to the Rose Group, for its chief substance, citral, is the first product of geraniol upon oxidisation. The two scents are often found together in leaves, as in those of Lemon Thyme. With other compounds, citral is present in the flowers of *Nymphaea, Magnolia* and in *Romneya coulteri* which are visited by beetles. A lemon-like fragrance is released by the leaves and stems of *Acorus calamus.*

(8) *Fruit-scented Group*

Though they could well be placed in the previous group, flowers having a fruit scent other than that of lemon (closely related to the rose perfume) are usually given a separate grouping. Several rambler roses have the scent of ripe apples, of oranges and bananas, and of our native flowers, those of the Gorse, *Ulex europaeus,* carry the refreshing scent of pineapple, with undertones of coconut or almond.

C

(9) *Animal Group*

The animal scents are closely related to the former group as in the Early Purple Orchid, *O. mascula*, which before fertilisation has a fruity scent with undertones of vanilla, which changes to a wet fur-like smell as the flowers fade. Again, *Anacamptis pyramidalis* has a fruity scent by day which changes to a foxy smell at night. The animal smells are related to the fruity scents by the esters of fatty acids. This is more apparent in the leaves of *Hypericum hircinum* which sometimes smell of goats, at other times of apples.

Where the acids are present in greater concentration and without the fruity undertones, the smell is most unpleasant, like stale perspiration, as in the flowers of *Centranthus ruber*, the Ox-eye Daisy and the Lizard Orchid. Valeric acid is, indeed, present in perspiration and gives the unpleasant smell found in the flowers.

(10) *Musk and Honey Group*

The scents given off by the flowers and leaves of this group are closely related to those of the previous one. They have the animal scent of the musk deer of the Himalayas from which musk is obtained. The scent is to be observed in diluted form in flowers of *Adoxa moschatellina*, Sweet Sultan, the Musk Orchid, *Herminium monorchis* and the Musk Rose; but it is most obvious in the hairy leaves of a number of native plants, including those of the Musk Storksbill, Musk Thistle and Musk Mallow. It is released from Clary leaves when drying and in the leaves of the Wild Strawberry.

The honey scent may be described as being a diluted musk perfume and is found in the flowers of the Buddleia and Sweet Scabious, and in Meadowsweet, but with slightly fishy undertones. Included here may be the leaves of the Common Box which release a musky or honey-like smell when brushed against on a warm day.

Though a number of leaves may be placed in the ten flower-scent groups, the larger number can be divided into four main groups – Turpentine, Camphor, Menthol and Sulphur – while those containing coumarin cannot be placed under any of the flower or leaf groupings. With leaves, the scent is more persistent and increases with age, while the opposite is true of flowers. This is due to the evaporation of moisture in leaves, the essential oil being left behind in concentrated form.

A: *Turpentine Group*

Here the essential oil is borneol acetate which is present in quantity in the leaves of several native conifers. The turpentine smell is also to be detected in the Fragrant Agrimony, *Agrimonia odorata*, due to the glandular hairs. It is also present in the leaves of Rosemary.

B: *Camphor and Eucalyptus Group*

The essential oil is eucalyptol which is present in Wild Marjoram,

Tansy, Thyme, Yarrow, Chamomile and Wormwood, producing a pungent but herby smell. Camphor is present in concentrated form in the leaves of Lavender, Sage and *Santolina incana*. In the leaves of Bay Laurel it is combined with a rose scent and with undertones of cloves to produce one of the most delicious of all leaf scents.

C: *Menthol Group*

The essential oil is menthol which produces the sensation of cold, an almost numbing sensation when inhaled in concentrated form. It is present in all forms of Mint and is particularly pronounced in Peppermint.

D: *Sulphur Group*

The pungent onion-like smell is due to compounds of sulphur which are absent from flowers. Several of the crucifers, e.g. Mustard, Watercress and the wild Garlics release the smell when bruised, the Garlics merely by wind movement.

BUTTERCUP FAMILY Ranunculaceae

Several acrid, poisonous plants due to presence of alkaloids. Mostly herbs or woody climbers present in hedgerows or marshlands. Leaves usually spirally arranged and often palmately lobed. Flowers borne solitary or in terminal inflorescence. With several species an involucre of leaves appears below the flowers. In *Aquilegia*, honey is secreted in a long spur; likewise in *Delphinium* (Larkspur) which usually bears blue or purple-blue flowers, pollination is by bees. The flowers are scentless and bees are attracted by their colour only. In those flowers where the honey is situated deep down, self-fertilisation is prevented by well-marked protandry, but is possible where the honey or pollen is more readily accessible, as in *Anemone* and *Ranunculus*; hence perfume is rare in the family.

HELLEBORUS Woody perennial herbs with leaves palmately or predately lobed. Flowers borne in terminal inflorescence; sepals 5, petals 5–10, tubular. Fruit 3–10 sessile follicles. It takes its name from Greek *helein*, to harm, and *bora*, food, a reference to its harmful properties.

STINKING HELLEBORE
Helleborus foetidus
Bears its flowers on a 3-ft stem. These are bell-shaped, yellow-ish green in colour, tipped with purple and are borne in terminal clusters. Trimethylamine is present which gives off an un-pleasant smell to attract midges and bluebottles for their pollin-ation. The lower stem leaves are evergreen with long stalks and are divided into 3–9 narrow lanceolate segments, but not to a common centre as with other species. Compounds of sulphur are present and the whole plant emits a most unpleasant smell, especially when handled, hence its country name. In bloom March–May.
Habitat: Deciduous woodlands

Stinking Hellebore

with a calcareous soil but mostly the Cotswolds, extending northwards into Northamptonshire. Rare in N England, Scotland and Ireland but present in S Wales.

History: In early times, the plant was thought to have powers of guarding the homestead from ills and was believed to be an antidote against madness. In the frontispiece to his *Anatomy of Melancholy*, Burton rather surprisingly coupled it with Borage 'to cheer the heart':

> Borage and hellebore fill two scenes,
> Sovereign plants to purge the veins
> Of melancholy, and cheer the heart
> Of those black fumes which make it smart.

CLEMATIS Climbing herbs or shrubs with opposite leaves, usually compound, and ending in a tendril, or with a twining petiole. Flowers usually hermaphrodite, borne solitary or in cymose inflorescence. Hairy style often persists on fruit. It takes its name from the Greek *klema*, a vine-shoot, a reference to its mode of climbing.

TRAVELLER'S JOY
Clematis vitalba

Traveller's Joy

A perennial herb with woody stems and opposite pinnate leaves, climbing by means of their twisting petioles. The clusters of greenish-white flowers, pubescent on the outside, are borne from the axils of the leaf stems and are sweetly vanilla-scented. They are without petals but have 4 sepals and numerous stamens and carpels. The fruit is an achene with conspicuously long feathery styles which persist well into winter and which assist in the dispersal of the seeds. From the whitish plumes, it takes its country name of *Old Man's Beard*. In bloom July–September.

Habitat: Hedgerows and woodland margins, usually in chalk or limestone districts, especially Sussex and Kent; also Derbyshire and Yorkshire.

History: Through the ages, it was a plant much loved by those travelling on foot or on horseback. Gerard (p. 19) in his *Herbal* (1597) said that it was common in every hedgerow from Gravesend to Canterbury 'making a goodly show', and beneath it weary travellers could rest; 'thereupon have I named it Traveller's Joy'. In an early eleventh-century Anglo-Saxon *Vocabulary,* it is called *Viticella-Woodebinde,* a name now used for the Honeysuckle.

AQUILEGIA Erect perennial herbs with ternately divided leaves. Flowers borne singly or in panicles; petals 5, growing downwards between the sepals to form a spur. Flowers blue, purple-brown, pink or white, the latter having a soft clove-like perfume. This is most pronounced in the subspecies *A. pyrenaica,* native of the Pyrenees but naturalised in Britain since earliest times. It also bears white flowers. Scent is absent in the blue and purple varieties.

COLUMBINE
Aquilegia vulgaris
subspecies *A. pyrenaica*

A valuable perennial of garden stature, happy in partial shade. The flowers are borne on 2-ft stems and arise from the axils of the leaves. They have a slightly drooping habit and in the garden form the spur is elongated and the flower several times larger than in the wild form. In the highly scented form *pyrenaica,* the spur is extremely long, slender and slightly incurved and is fertilised by long-tongued moths; *A. vulgaris* is visited by bees. The flowers are of bluish-white colouring. The attractive trifoliolate leaves have a greyish appearance. In bloom May–July.

Habitat: Deciduous woodlands and other moist places, especially hedge-rows and fenlands throughout

Columbine

England and Wales. *A. pyrenaica* is naturalised in parts of Scotland, especially Angus and in N England.

History: It is a plant of the cooler regions, enjoying a shady situation and a soil retentive of moisture when it will seed itself and prove to be fully perennial. Shakespeare coupled it with Mint when in *Love's Labour's Lost* he wrote:

> Armado: I am that flower.
> Dumain: That mint.
> Longaville: That columbine.

The reason for this is possibly that both plants have a liking for moist conditions and the playwright may well have seen the two plants growing in close proximity. Or maybe it was because both plants possessed similar medicinal properties. In Shakespeare's time, hot mint cordial was used to cure a sore throat, while John Guillam stated in *A Display of Heraldry* (1632) that the Columbine 'is holden to be very medicinable for the dissolving of swellings in the throat'. The flower was so widely used to cure all manner of aches and pains that it came to be associated with 'folly'. In his play *All Fools* (1605) George Chapman, translator of Homer, wrote:

> What's that – a columbine?
> No. That thankless flower grows not in my garden.

During the time of Spenser and Shakespeare it was also associated, like Fennel, with flattery. Spenser compared his wife's neck 'like unto a bunch of cullambynes' and in *Hamlet* Ophelia says: 'There's fennel for you, and columbines'.

Its country name is derived from the Latin *columba*, a dove, for the nectaries, like long spurs, are thought to resemble the head and neck of a dove. There are five nectaries to each bloom which the French artists of the fourteenth century took to represent the Seven Gifts of the Holy Spirit, correcting this by painting the plant with seven blooms. Its botanical name is derived either from the Latin *aquilegus*, a water carrier, an allusion to the ability of the flowers to hold water, or from *aquila*, an eagle, whose claws the flowers resemble.

The poet Clare described the colour of the wild Columbines as 'stone-blew or deep night-brown' which is a most accurate description of the flowers of *A. vulgaris*. Donald McDonald likened their perfume to that of newly mown hay, but it is more spicy in the white form and is absent from the wild 'blue'.

BARBERRY FAMILY Berberidaceae

Woody perennial (mostly) evergreen plants of northern temperate regions with alternate, usually spiny, pinnate leaves, simple or compound. Flowers hermaphrodite, borne solitary or in cymes or racemes; petals equal in number to sepals, or twice as many. Stamens equal to petals. Fruit, a berry.

BERBERIS Evergreen or deciduous shrubs with long and 'short' shoots showing a close relation to Coniferae. Leaves simple, usually spiny. Flowers borne in racemes, afterwards growing out to long shoots.

BARBERRY
Berberis vulgaris
Deciduous shrub attaining 6 ft in height
with yellow bark and sharp spines on
the twigs. The oval leaves are also sharply
toothed, while the bright lemon-yellow
flowers are borne in drooping clusters.
They have a delicate scent and are
visited by flies and bees which suck the
honey flowing into the angles formed
between the stamens and ovary. In
bloom May–June.
Habitat: It is found in deciduous
woodlands and hedgerows and on
waste ground, scattered throughout the
British Isles but is becoming scarce. Introduced into Ireland at an early date.

Barberry

History: From the oblong bright red fruits, a preserve was made while
the spiny plants were used to enclose fields, until it was found that they
harboured a parasitic fungus which lived part of its life on the leaves
and partly in wheat and caused 'rust' disease.

MAHONIA Evergreen shrubs with pinnate leaves, *M. aquifolium*
being introduced from North America in the seventeenth century when
it was widely planted for game coverts and shelter-belts. Flowers
yellow-orange, borne in racemes.

OREGON GRAPE   *Mahonia aquifolium*
Native of Oregon and so-named because of its black grape-like fruit,
covered in 'bloom'. An evergreen with ovate pinnate leaves and
growing 2–3 ft tall. Stems spineless. The flowers are deepest yellow and
are borne in a many-sided panicle. They are heavily scented like Lily of
the Valley and, like most winter flowers, are pollinated by the various
forms of insect life about at this time. In bloom March–May.
Habitat: Naturalised in deciduous woodlands, usually growing in
calcareous soils. It is now more common than the native *Berberis
vulgaris*.

WATER-LILY FAMILY   Nymphaeaceae

Perennial water-plants with a short or creeping rootstock. Leaves
orbicular, leathery, with basal sinus, usually floating. Sepals 3–6,
passing into petals, these into stamens inserted on a fleshy disc which

surrounds a many-chambered ovary. The starch-containing roots were at one time used as food in times of famine.

NYMPHAEA Shallow water perennials with creeping rhizomatous rootstock from which leaves and flowers rise on long stems which float on the surface of the water. Sepals 4, green on outside; petals white, numerous, stamens many, adnate to the disc. Fruit, a berry-like capsule which ripens underwater. The plant was so named when it was found growing in places nymphs were also believed to inhabit.

WHITE WATER-LILY *Nymphaea alba* Pl. 1

It has large circular deep green leaves, often reddish-brown on the underside and with deep basal lobes. The flowers are purest white, up to 8 in. across with golden stamens and open towards midday, closing and partially submerging towards evening. They emit a sweet fruity perfume when open in sunlight and are visited by Cetoniae (beetles). In bloom July–August.

Habitat: Present in lakes, ponds and occasionally in marshlands throughout the British Isles; especially prominent in SW England and in Ireland. The variety *N. alba minor*, with smaller flowers, is represented in Scotland.

NUPHAR Perennial with a stout rhizomatous rootstock and leaves without stipules, floating or partially submerged. The flowers yellow, globose and rising above the water. Sepals 5–6; petals shorter than sepals. Stamens numerous, hypogynous with broad filaments. Fruit, berry-like ripening above the water.

YELLOW WATER-LILY *Nuphar lutea* Pl. 1

From the brandy-like smell of its globular yellow flowers, it is also known as *Brandy-bottle* and is visited by beetles of the genus *Meligethes*, attracted to its smell which is similar to that of ripe plums. This is due to the combination of acetic acid and ethyl alcohol to form ethylacetate. The leathery leaves are submerged or floating and are ovate with a basal sinus. The flowers are small, about 3–4 in. across, with spathulate petals shorter than the concave sepals. The stigmatic disc has 15–20 rays. The fruit is flask-shaped and berry-like, bright green in colour. In bloom July–September.

Habitat: Frequenting slow-running water, lakes and ponds throughout England and Wales but uncommon in Scotland where it is replaced by *N. pumila*, the Lesser Yellow Water-lily, which has only slightly scented flowers.

History: Robert Burton in his *Anatomy of the Melancholy* said that Laurentinus suggested bringing scented Water-lilies indoors, 'to evaporate in the room and which will make a more delightful perfume if there be added orange-flowers; pills of citron; rosemary; cloves; bays; labdanum; styrax and such-like gums . . .'.

CABBAGE FAMILY Cruciferae

Annual or perennial herbs, rarely shrubs, inhabiting northern temperate zone, with radical leaves formed in rosette fashion and often containing a pungent, watery juice. Flowers with 4 petals, formed crosswise, usually white, yellow or pale mauve; alternating with 4 sepals. Stamens 6, of which 2 are honey-secreting glands. Many genera of the family are valuable garden vegetables. Sulphur compounds are present in the foliage, e.g. *Nasturtium* (Water Cress), *Brassica, Diplotaxis* (Stinkweed). Few bear flowers with scented attractions; they are usually inconspicuous and attract few insect visitors. The flowers are mostly self-pollinating. In cold wet weather, the flowers may only partially expand when the anthers of the longer stamens remain in contact with the stigma (as in *Rorippa sylvestris*) and self-fertilisation is effected.

DIPLOTAXIS Annual or perennial herbs differing from *Brassica* in the arrangement of the seeds in the pods. These are in a double instead of a single row, hence its name from two Greek words, *diplos,* double and *taxis,* rank. Slender, branched plants with pinnately cut leaves and stems sparsely covered in stiff hairs. Leaves release a pungent fetid smell when pressed. Flowers slightly scented.

STINKWEED *Diplotaxis muralis*
It is a native of central and southern Europe, but became naturalised early in our history, mostly in southern England. An annual or biennial branched at the base and with erect hairy stems up to 12 in. tall. The leaves are toothed or have triangular lobes and are produced in a basal rosette. They release an unpleasant fetid smell when bruised, like sulphurated hydrogen, hence its name. The flowers are pale yellow and are slightly scented. Pollination is mostly by bees (Apidae) though in default, the upper part of the long anthers bends back to touch the stigma and so effects self-fertilisation. In bloom May–September.
Habitat: Waste ground, especially of a sandy nature and mostly confined to S England, usually near the coast.

WALL ROCKET *D. tenuifolia*
Perennial and much-branched plant growing up to 2 ft high with a woody stem and glaucous leaves divided into narrow segments, which release a fetid smell when handled. The pale yellow flowers are more

fragrant than those of *D. muralis*, with
the petals twice as long as the spreading
outer sepals; the stigma broader than
the style. Pollination is mostly by bees,
in the absence of which self-pollination
takes place. In bloom May–August.
Habitat: Naturalised on old walls,
hence its name Perennial Wall Rocket.
Especially common in S England. Also
present on banks and waste ground of a
chalky nature.

CORONOPUS   Cosmopolitan weeds,
annual or perennial, of prostrate habit,
usually with pinnatifid leaves. Flowers
inconspicuous with small petals or
none; stamens 6, 4 or 2, without
appendages. Seed-pods flattened.

SLENDER WART CRESS
Coronopus didymus
It is an almost prostrate annual with
hairy stems and pinnate leaves which
release a sulphur-like smell, especially
pronounced in warm weather, when
handled or if trodden upon. The basal

Wall Rocket

leaves are stalked. The flower petals are white and are shorter than the
sepals or may be absent altogether. 2 fertile stamens are usually present
and the flowers are self-pollinating. In bloom June–September.
Habitat: Weeds of wayside and waste ground, especially common in
S England but extending to N Scotland; also in Ireland.

ALYSSUM   Annual or perennial plants of dwarf spreading habit,
several of which are naturalised garden escapes. The simple linear
leaves are covered with stellate hairs, while the flowers are borne in
racemes without bracts. Stamens 6, usually with appendages.

SWEET ALISON   *Alyssum maritimum*   Pl. 2
So named because its flowers have the refreshing scent of newly mown
hay. It is a dwarf spreading plant with linear lance-shaped downy leaves
1 in. long. It bears mounds of small white flowers in dainty racemes.
In bloom June–September.
Habitat: Wastelands, usually near the coast of S England; also on old
walls.

MATTHIOLA Branching annual, biennial or perennial herbs, mostly inhabiting S Europe and N Africa. Growing 10–18 in. tall, the leaves and stems are grey with hairs; the leaves entire, sinuate or pinnatifid. The flowers borne in racemes; usually pale mauve or flesh-coloured; the sepals erect; petals with claws. Stigma of 2 horned lobes. Flowers powerfully scented, pollination being by butterflies by day; *M. sinuata* visited by moths in evening.

HOARY STOCK
Matthiola incana

Hoary Stock

The parent of the garden Stocks. It is a woody plant growing 10–15 in. tall covered in glandular hairs. The narrow lanceolate leaves are formed in rosettes, while the flowers are borne in loose racemes. The flowers contain several essential oils, present in leaves and barks, such as eugenol and cinnamic compounds, which provide the aromatic spicy perfume, more pronounced in the Double Stocks. It is a more pleasing scent than that of the heavily-scented flowers such as *Daphne* and *Viburnum*, which contain methyl indol, the active principle of civet which has a disturbing effect upon some people. The flowers measure nearly 2 in. across and are to be found in shades of pale mauve and violet (also white), attracting mostly butterflies for their pollination. The sepals are covered in tiny hairs. In bloom May–July.

Habitat: M. incana was originally found on the limestone cliffs of the southern extremities of the Isle of Wight, but from earliest times it has occurred on the chalky cliffs of Sussex and Dorset and in several other places near the sea in S England; also N Yorkshire and Durham.

History: The Hoary Stock was a common garden plant during Elizabethan times for Gerard tells us that it was 'greatly esteemed for the beauty of its flowers and pleasant sweet perfume'. He mentioned that Stock Gillofloures 'do grow in most gardens in England'. It was described by Dr Turner in his *Herbal* (1568) as the *Stock Gelouer*; also by Lyte in his *New Herbal* (1578), to distinguish it from the Wall Gillofer (Wallflower) and the Dame's Gillofer (*Hesperis matronalis*), each of the same family and bearing flowers with the familiar clove perfume of the Spice Islands. The poet Spenser mentions 'gilloflowers' (p. 20) and James Thomson in *The Seasons* mentioned 'the . . . lavish Stock, that scents the garden round'.

Henry Phillips in the *Flora Historica* brings to our notice the enduring quality of the Stock's perfume. He wrote, 'though less graceful than the rose and not so superb as the lily, its splendour is more durable, its fragrance of longer continuance'. Stocks bridge the gap between the early spring flowers and those of midsummer, flowering from the end of April until early July and, like the Wallflower, emit so strong a fragrance when warmed by the sun that their presence may be detected from a distance. John Clare, the Northamptonshire ploughman-poet, writing from his cottage at Helpston, mentioned their long-flowering qualities:

> The white and purple gillyflowers, that stay
> Lingering in blossom, with summer half away.

From *M. incana* Brompton Stocks were raised at the Brompton Road nursery of Messrs London and Wise (who laid out the gardens at Blenheim Palace) early in the eighteenth century. This strain of biennial plants bloom at much the same time of year.

GREAT SEA STOCK *M. sinuata* Pl. 2

So called for its liking for a situation close to the sea. It resembles *M. incana* except that it is more bushy, more hairy, and has a greyish appearance. The basal leaves are sinuate or pinnatifid and covered in small black spots. The pale mauve flowers are of 1-in. diameter and are borne in a loose raceme. They are much visited by Lepidoptera. In bloom June–September.

Habitat: Occasionally on sand-dunes and rocky limestone formations near the sea in Kent and Sussex; N Devon; Pembrokeshire; and Co. Clare in Ireland.

HESPERIS Erect biennial or perennial hairy herbs with ovate or oblong, entire, toothed leaves and stem branched and hairy. Flowers large, borne in bractless racemes, white or mauve. Sepals erect, petals clawed. Stamens 6; style short; stigma 2-lobed. Seed-pods cylindrical; seeds in single row in each cell.

DAME'S VIOLET, SWEET ROCKET, DAME'S GILLYFLOWER *Hesperis matronalis* Pl. 2

Grows wild from C Europe to SE Russia and was introduced into Britain at an early date. A hairy perennial growing 3 ft tall, it has toothed lanceolate leaves, tapering at the base. The flowers are borne in loose racemes, each one measuring about 1 in. across and possessing a powerful scent, especially at night. This basically resembles the violet perfume, but is spicy with undertones of cinnamon, for cinnamic alcohol is present. The flowers are usually white (occasionally mauve or purple) and are visited by night-flying Lepidoptera, though they are also capable of self-pollination, with the anthers bending inwards to cover the stigma with pollen. In bloom mid-May–mid-July.

Habitat: As it is a garden escape, it is usually found in hedgerows and grass verges near villages. It is also seen growing between the stones of old walls like Wallflowers and Stocks which bloom at about the same time. Also on cliffs close to the sea.

History: H. matronalis has grown in English gardens from earliest times. Gerard described it as having 'large leaves of a dark green colour, snipt about the edges; . . . the flowers come forth at the top of the branches like those of the Stock Gillofleur, of a very sweet smell'. During Elizabethan times they were known as the *Queen's Gilloflowers* for they were considered superior in perfume to all other flowers. A double white form, *flore plena,* was known as the *Double White Rocket* or *Double Garden Rocket.* It is propagated only by cuttings, for it does not set seed and its roots are almost impossible to divide. The flower is illustrated in Parkinson's *Paradisus* (1629) under *Hesperis pannonica.*

CHEIRANTHUS Biennial or perennial herbs or undershrubs, native of S Europe, Madeira and N Africa and possibly introduced into Britain by the Romans. They form much-branched plants, growing 9–18 in. tall with oblong entire or toothed leaves covered with appressed hairs. The flowers are borne in loose racemes without bracts. The individual flowers are large, measuring more than 1 in. across and have long-clawed petals. Stamens 6, with nectaries at the base of the 2 outer stamens. Style short; stigma with 2 spreading lobes. Seeds arranged in 1–2 rows in cylindrical pod. In bloom April–June.

WALLFLOWER *Cheiranthus cheiri* Pl. 2

Very familiar and one of the most loved of all flowers for garden display in spring and early summer. It is a perennial herb with a long taproot and an erect much-branched woody stem. The 2–3-in. long lance-shaped leaves are entire and are borne along the angular stems. The flowers appear in long racemes and are orange-yellow in the wild state, occasionally brown and with a pronounced violet-like perfume with spicy undertones. The scent of the Wallflower is highly complex for the main elements of the Violet, Rose, Hawthorn and Orange blossom have been isolated from its attar. The violet element is most pronounced in the double forms and tends to fade after being inhaled for a little time, to be replaced by a more spicy element which is even more pronounced in *Hesperis matronalis*, the Stock and the Clove-scented Carnation and Pink. The essential oil of these flowers contains eugenol and cinnamic alcohol, which provide the pleasant spicy undertones. The flowers are much visited by bees, which are directed to the nectar by the orange-yellow veining conspicuous in the yellow flowers. For garden display it is usually treated as a biennial, the seed being sown in May, for the plants require a long period to mature before planting in beds in November. In bloom April–June.

Habitat: Throughout the British Isles, usually high up on old walls and cliffs, out of reach of passers-by.

History: To the early English it was known as *chevisaunce*, the comforter, because of its warm comforting scent. Edmund Spenser coupled it with the Pansy in his *Lycidas*:

> The pretty paunce and the chevisaunce
> Shall match with the faire flower-de-luce.

The plant may have reached Britain with the Roman invasion or may have been introduced at a later date by the Normans, together with the Pink. Their flowers were used to flavour drinks, and both plants may be seen growing together about the walls of ancient castles. Both love an open, sunny situation and flourish where their roots can penetrate into mortar. *Cheiranthus cheiri* was planted about the walls of castles and manors, especially near windows so that the scent could enter and counteract the musty atmosphere. The Wallflower is to be found on the walls of Neidpath Castle, situated on the banks of the Tweed near Peebles, and legend has it that while the Earl of March occupied the castle, his younger daughter Elizabeth was betrothed against her will to the son of King Robert of Scotland, though she was in love with the son of a border chieftain. One day her lover, disguised as a minstrel, sang beneath the window of her room. The words of his song suggested that if it was Elizabeth's wish to elope with him she should reach for and drop a Wallflower growing near her window, at his feet; but in so doing, she herself fell to the ground, dying in her lover's arms. Ever after he wore a sprig of Wallflower when it was in bloom and from that time the flower became the symbol of fidelity.

Sir Walter Scott, who must have often come upon the tower of Neidpath from his home at nearby Abbotsford, may have had the legend in mind when he wrote these lines:

> And well the lonely infant knew,
> Recesses where the wall-flower grew,
> And honeysuckle loved to crawl
> Up the low crag and ruin'd wall . . .

It was the favourite flower of the Border Minstrel and he leaves us in no doubt as to his affection for it:

The rude stone-face, with wall-flowers gay,
To me more pleasure yield
Than all the pomp imperial domes display.

Bernard Barton, the Quaker poet, expressed similar feelings in his lines written at Leiston Abbey:

And where my favourite abbey nears on high
Its crumbling ruins and on their loftiest crest,
Ye wallflowers, shed your tints of golden dye,
On which the morning sunbeams love to rest.

But only John Clare mentioned its delicious scent: 'The single blood walls of a luscious smell'.

Parkinson described six kinds in the *Paradisus*, including the double yellow. He wrote that 'the sweetness of the flowers causeth them to be used in nosegays', hence it came to be given its botanical name *Cheiranthus*, or *hand-flower*, for whenever the plant was in flower it was carried in the hand by ladies on special occasions such as weddings and church festivals. Parkinson and William Lawson, both writing at the same time, referred to it as the *Wall Gilloflower* while to Gerard it was known as the *Yellow Stock Gilloflower*, to distinguish it from the mauve *Matthiola incana*. John Rea of Bewdley in his *Flora, Ceres and Pomona* (1665) described the flower as being 'as broad as a half-crown' for by then the quality had been much improved. Further improvement was achieved by Robert Ingram, head gardener to the Duke of Rutland at Belvoir Castle, the name he gave to his special strain of the Yellow Wallflower which to this day remains outstanding among many lovely strains.

In his essay *Of Gardens*, Francis Bacon tells us that they (Wall-flowers) were 'very delightful to be set under a parlor or lower chamber window', and Tournefort said that 'the distilled water of the flowers cleanses the blood and begets a cheerful disposition'.

ALLIARIA Annual or perennial herbs of erect habit; the stems covered in hairs; the leaves cordate and long-stalked. The flowers white, petals short-clawed; stamens 6, without appendages; style short; stigma capitate.

JACK-BY-THE-HEDGE, HEDGE GARLIC, GARLIC MUSTARD
Alliaria petiolata

A biennial herb with all its parts, including the taproot, smelling of garlic, hence its botanical name. It forms an erect unbranched hairy stem with long-stalked pale green basal leaves, the stem leaves being short-stalked with toothed margin. The flowers are white and appear in small

racemes. They have an unpleasant smell
and attract midges for their pollination,
though in the absence of these, self-pollination takes place. In bloom April–May.
Habitat: Common in hedgerows and at the
edges of woodlands throughout the British
Isles but mostly where the soil is of a
chalky nature and shaded from the direct
rays of the sun.

VIOLET FAMILY Violaceae

Annual or perennial herbs or shrubs with
alternate (rarely opposite) leaves, simple,
entire. Flowers hermaphrodite, axillary,
regular or irregular, solitary or in panicled
cymes; 2-bracteolate. Sepals 5, elongated
into a spur. Petals 5, hypogynous, equal or
unequal. Stamens 5, alternate with petals;

Garlic Mustard

filaments short. Fruit, a capsule or berry; endosperm fleshy. A single
family connected to no other.

VIOLA   Cosmopolitan, inhabiting mainly the cooler regions of the
world. Annual or perennial herbs with alternate leaves and leafy
stipules. Flowers borne mostly solitary, axillary; sepals equal; petals
spreading, lower ones spurred. Stamens 5, alternate with petals, two
lower ones spurred. Style straight or curved. Fruit, a 3-valved capsule.

SWEET VIOLET *Viola odorata*
The only scented British species. A
perennial, it has a short rootstock with
long runners, and leaves which are
heart-shaped at the base; stipules
glandular. The white or purple-blue
flowers are spurred with 4 upper oblong
petals, the 2 side-petals with or without
a tuft of hairs. They are visited by
Hymenoptera which insert their probosces beneath the stigma and push up
the pistil, opening the conical ring of
anthers. The pollen is thus shed upon
the proboscis and in every flower the
insect first touches the stigma, and then

Sweet Violet

causes shedding of the pollen. In bloom March–June.

Habitat: Moist grassy banks, in hedgerows and deciduous woodlands throughout the British Isles.

History: Long before the birth of Christ, *V. odorata* was in commercial cultivation for its sweetening properties and was so highly regarded by the Greeks that the flower became the symbol of ancient Athens. In France, the flower has been held in equal regard from earliest times and shortly before his exile, Napoleon picked blooms from the grave of his beloved Josephine, which were later found in a locket on his death-bed.

In England, the blooms were cooked with meat and game and were grown in every garden for this purpose. In the library of Trinity College, Cambridge, is a fifteenth-century treatise written in rhyme by one 'Master John, Gardener' in which are mentioned several flowers grown for their sweetening powers:

> Peruynke, violet, cowslippe and lyly,
> Holyhocks, coryawnder, pyony.

Violet blooms were also in great demand for strewing on the floors of church, manor house and cottage to relieve the damp, musty smell. Thomas Tusser (p. 19) mentions the Violet in a list of twenty-one herbs and flowers all of which possess fragrance either in their flowers or foliage, and which were suitable for strewing on floors. In his *Five Hundred Points of Good Husbandry* he also mentioned the Violet, among other flowers, as being suitable for 'windows and pots' (p. 20). Indeed these richly perfumed flowers were widely grown in earthenware pots to be placed on the inside window-ledges of houses where their fragrance was much appreciated.

In a similar way, the Violet was widely used during ancient times as a cure for headaches, to relieve melancholia and insomnia, the blooms being placed upon the head, especially at night. An interesting cure for sleeplessness is given by Anthony Ascham in his *Little Herbal* of 1525. He wrote, 'for they that may not sleep, seep this herb in water and at even, let him soak well his feet in the water to the ancles, and when he goeth to bed, bind of this herb to the temples and he shall sleep well by the Grace of God'. Syrup of Violets was also widely used as a safe purgative for small children, and Gerard advised adding to an ounce of the syrup 'eight or nine drops of oile of vitriol' (sulphuric acid) and he added that 'sucking children may take it without peril'.

Lending itself admirably to cultivation in the small rectangular beds of medieval gardens, the plant was to be found in those of early monasteries and in all gardens until the mid-seventeenth century, when landscape gardening first became prominent. Gerard wrote of the Violet in a charming manner when he said, '. . . gardens themselves receive by these the greatest ornament of all chiefest beautie and most gallent grace, and the recreation of the minde, which is taken thereby cannot but be very good and honest; for they admonish and stir up a man to that which is comely and honest for flowers, through their beauty and exquisite form do bring to a gentlemanly minde the remembrance of honestie, comeliness and all kindes of vertues. For it would be an unseemly and filthie thing for him that doth looke upon and handle faire and beautiful things and who frequenteth faire and beautiful places, to have his minde not faire, but filthie and deformed'.

No garden flower possesses greater charm or a more delicious fragrance. In his essay *Of Gardens*, Francis Bacon described the Violet as possessing the sweetest scent of all flowers. From the time of the sixth century poet-bishop of France, Fortunatus, poets everywhere have honoured the Violet above all other flowers, but it was Shakespeare who gave it most prominence, alluding to the plant on eighteen different occasions. What is more, his allusions leave us in no doubt as to which flower he really meant, and for the first time the name of the Sweet Violet, now so familiar to us, became fixed.

Before that time, almost all flowers with a similar fragrance were named *Viola*. The Sweet William was *V. barbata*, the Sweet Rocket, *V. matronalis*. But his accurate descriptions of the flower put Shakespeare's own seal upon it. It is forever Shakespeare's Violet, as in *Sonnet xcix*:

> The forward violet thus did I chide –
> Sweet thief, whence didst thou steal thy sweet that smells,
> If not from my love's breath? The purple pride
> Which on thy soft cheek for complexion dwells,
> In my love's veins thou hast too grossly dy'd.

Nothing about the flower escaped the eye of the poet for he must have looked upon the springtime Violet many times, catching its every mood, in sunshine and shower, on his rambles round the countryside of his native Warwickshire. Angelo tells us in *Measure for Measure*:

> The tempter or the tempted, who sins most? Ha!
> Not she; nor doth she tempt: but it is I,
> That, lying by the violet in the sun,
> Do, as the carrion does, not as the flower,
> Corrupt with virtuous season.

'Lying by the violet in the sun' must have occupied many of Shakespeare's happiest moments, as he relaxed upon the grassy bank leading from the garden of Anne Hathaway's cottage to the old orchard, carpeted with purple Violets hiding their faces in the grass. Perhaps this was the very bank he had in mind when he wrote in *A Midsummer Night's Dream*:

> I know a bank whereon the wild thyme blows,
> Where oxlips and the nodding violet grows.

1

2

Plate 1
Water-lily Family   1 Yellow Water-lily   2 White Water-lily

Plate 2
Cabbage Family 1 Sweet Alison   2 Wallflower
3 Great Sea Stock   4 Dame's Violet

Shakespeare loved the Violet, possibly before all other flowers, for its fragrance and because it was one of the earliest of wild flowers to bloom, heralding the approach of summer. He shows his pleasure in the springtime Violet when in *Richard II* the Duke of Aumerle, Richard's cousin, is welcomed by his parents, the Duke of York and his Duchess, at their palace:

> Duchess: Welcome, my son: Who are the violets now
> That strew the green lap of the new-come spring?
> Who are the courtiers now attending on
> Bolingbroke in the spring-time of his reign?

In *Cymbeline*, the Queen of Britain commands her ladies-in-waiting:

> Whiles yet the dew's on ground, gather
> Those flowers; make haste . . .
> The violets, cowslips, and the primroses,
> Bear to my closet.

Shakespeare concludes *Love's Labour's Lost* with Spring singing a delightful ditty which expresses the poet's pleasure at the lengthening days and the meadows painted with hues of springtime:

> When daisies pied and violets blue,
> And Lady-smocks all silver white,
> And cuckoo-buds of yellow hue,
> Do paint the meadows with delight . . .

In *A Winter's Tale*, the poet, still grieving for his dead son, wove the flower into the love-lorn Perdita's garland of spring flowers in the most lovely of all descriptions of springtime:

> Perdita: I would I had some flowers o' the spring . . . daffodils,
> That come before the swallow dares, and take
> The winds of March with beauty; violets dim,
> But sweeter than the lids of Juno's eyes,
> Or Cytherea's breath . . . O! these I lack,
> To make you garlands of; and, my sweet friend,
> To strew him o'er and o'er.

Yet again, in the same passage, Shakespeare tells us of his delight in lying relaxed upon a grassy bank of Violets.

With their sombre colouring and the humility with which the flowers appear, almost hidden by the vegetation around them, Violets have always been associated with death and funerals:

> The purple violets and marigolds,
> Shall as a carpet hang upon thy grave
> While summer days do last.

Marina in *Pericles* says this as she enters with her basketful of mourning tributes on the death of her mistress. Again, in *Hamlet*, the distracted Ophelia, speaking of the death of her father Polonius, says to Laertes, her brother:

> There's a daisy. I would give you some violets
> but they withered all when my father died.

Later, Laertes mourns the death of Ophelia:

> Lay her i' the earth;
> And from her fair and unpolluted flesh
> May violets spring!

In *Sonnet xii*, Shakespeare compares the fading beauty of the Violet with a person of handsome features in lines of great tenderness:

> When I behold the violet past prime,
> And sable curls all silver'd o'er with white . . .
> Then of thy beauty do I question make,
> That thou among the wastes of time must go,
> Since sweets and beauties do themselves forsake
> And die as fast as they see others grow.

Compare these lines with those of Henry Vaughan in *Daphnis* (1678) in which he tells of his sadness at the short season enjoyed by the Violet in bloom:

> So violets, so doth the primrose fall
> At once the spring's pride and its funeral,
> Such early sweets get off in their still prime,
> And stay not here to wear the foil of time.

Shakespeare confirms his intimate knowledge of the Violet in his numerous passages devoted to its perfume, as in *Venus and Adonis* when Venus singles out the flower as possessing the most delicious of all flower scents, comparing it to that of Adonis and coupling it with his death:

> Who when he lived, his breath and beauty set
> Gloss on the rose, smell to the violet.

Its incomparable delicacy is introduced to us in *King John* when Salisbury, in reference to the twice crowning of the King, addresses him with these touching words:

> To gild refined gold, to paint the lily,
> To throw a perfume on the violet,
> To smooth the ice, or add another hue
> Unto the rainbow, or with taper-light
> To seek the beauteous eye of heaven to garnish
> Is wasteful and ridiculous excess.

Moreover, he carefully noted the colour and veining of the blooms, describing them in *Venus and Adonis* as 'These blue-veined violets' for the streaks that radiate from the centre of the bloom, as in *Viola tricolor*, are plainly seen upon close inspection of the flower. In *A Winter's Tale* he alluded to this same veining to be seen on the eyelids as a faint blue network where the skin is white and translucent.

Nearly always Shakespeare refers to a 'bank of violets', as he does in the opening passages of *Twelfth Night*. Again, in *Hamlet*, Laertes tries to give comfort to his sister Ophelia on the departure of Hamlet:

> For Hamlet, and the trifling of his favour,
> Hold it a fashion, and a toy in blood;
> A violet in the youth of primy nature
> Forward, not permanent, sweet, not lasting,
> The perfume and suppliance of a minute;
> No more.

George Herbert, born during Shakespeare's lifetime, also wrote of the quickly fading scent of the Violet in lines written shortly before his death at the age of 39:

> Farewell, dear flowers, sweetly your time ye spent
> Fit, while ye lived, for smell or ornament,
> And after death for cures;
> I follow straight without complaint or grief,
> Since if my scent be good, I care not if
> It be as short as yours.

The reference to the fleeting fragrance of the bloom is interesting, for the Violet's perfume is strangely quick to fade. This is because its chemical composition contains a substance known as ionine (from the Greek *ion*) from which the word *violet* is derived. This substance is able to dull the sense of smell within a very short time and so it is not the flower which loses its fragrance when cut and inhaled but our own powers of smell which are dulled. If, after a few moments the bloom is again smelled, the perfume will return in the fullness of its strength, only to disappear again. It is this quality which has contributed so much to the charm of the Violet for one can never become intoxicated with an excess of its sweet fragrance.

The humility of the Violet has been a constant inspiration to the poets. Milton refers to it in *Paradise Lost*:

> . . . underfoot the Violet,
> Crocus and Hyacinth with rich inlay
> Broidered the ground, more colourful than with stone
> Of costliest emblem.

And St Francis de Salles said that 'a widow is, in the church, as a little March Violet shedding around an exquisite perfume by the fragrance

of her devotion; and always hidden under the ample leaves of her
lowliness and by her subdued colouring . . . she seeks untrodden and
solitary places . . .'

The poet Barton made the same allusions with his delightful words:

> Beautiful are you in your lowliness,
> Bright in your hues, delicious in your scent;
> Lovely your modest blossoms, downward bent
> As shrinking from your gaze . . .

Wordsworth in the poem 'She dwelt among the untrodden ways'
compares the sweet charm of a maiden, Lucy, to that of:

> A violet by a mossy stone
> Half-hidden from the eye! . . .

It was Queen Victoria who set a new fashion in Violets during her
reign, using the bloom in posies not only for evening wear, but for all
daytime occasions. During her latter years, it was said that over 4,000
plants were grown under frames at Windsor, to provide the Queen and
the ladies of the Court with a constant supply of blooms.

Durandus, Bishop of Mende (1295), regarded the Violet as the
emblem of the humility of confessors.

Moreover, Bishop Fortunatus once wrote to Queen Radegonde when
she was at the nunnery at Poitiers saying, 'Of all the fragrant herbs,
none can compare in nobleness with the purple violet. They shine in
royal purple: perfume and beauty united in their petals'. That they will
surely be found 'in the starry meadows beyond Orion', was the wish of
Edgar Allan Poe.

In a letter of 13th April, 1819, sent by John Keats to his wife Fanny,
he writes 'I hope you have good store of double violets – I think they
are the princesses of flowers, and in a shower of rain almost as fine as
barley-sugar drops are to a schoolboy's tongue'. For the ladies to wear
a bunch of Violets was the height of fashion during Keats' time and
Double Violets were the most fragrant of all:

> Between her breasts, that never yet felt trouble,
> A bunch of violets, full blown and double,
> Serenely sleeps.

For centuries Violets have been sold in the streets of London. 'Buy
my sweet violets' was for long a familiar cry, though this is no longer
applicable to the modern posies usually made up of the scentless
Governor Herrick variety. The use of modern insecticides, however,
should once again bring about a return to perfumed varieties, which are
particularly troubled by pest and disease but are still grown in large
quantities to supply the perfume industry in the making of soaps and
scent, which like that of the Lavender always retains its popularity.

Surprisingly, almost all the Violets sold in the streets during the nineteenth century were imported from France, the home of the commercially cultivated Violet until quite recent times. They are now grown in Devon, Dorset and Cornwall to supply the commercial market during springtime.

ST JOHN'S WORT FAMILY *Hypericaceae*

Mostly of temperate regions; occasionally in mountainous regions of tropics. Trees, herbs or shrubs with red resinous juice. Leaves opposite, simple, exstipulate, often with resinous glands. Flowers yellow, borne in cymes or umbels. Sepals 5; petals 5. Stamens free, or cohering at base into 3–5 bundles; anthers versatile. Ovary superior; fruit, a dry capsule.

HYPERICUM Evergreen perennial herbs or shrubs; leaves opposite, sessile. Flowers without nectar. Visited by Hymenoptera, though petals and stamens draw together to bring anthers and stigma in contact, thus effecting self-fertilisation. The yellow flowers are scentless though the sessile leaves have transparent resinous dots (glands) and release a foxy fur-like smell when pressed, which is not unlike that of flowers of the Animal Group. The minute oil-glands may be observed by holding the leaves to the sunlight when the glands or cells, devoid of protoplasm and no larger than pin-pricks, are clearly visible. Caproic acid is present in the cells and it is this which emits the somewhat unpleasant foxy or goat-like smell. This is most pronounced in the species *H. hirsutum*. In *H. androsaemum*, the smell is pleasantly resinous and in *H. perforatum*, somewhat lemony.

SWEET AMBER *Hypericum androsaemum* Pl. 3
Also named *Tutsan*, from the French *toute-saine*, all-heal. A semi-evergreen undershrub, its red stems have 2 raised lines. It is a handsome perennial growing 2–3 ft tall with large oval leaves 3 in. long and camphoraceous, the scent being retained after drying. The golden-yellow flowers of 5 petals are about 1 in. across with attractive radiating stamens arranged in bunches of 5. The flowers are followed by fleshy purple-black fruits. In bloom July.
Habitat: Damp woodlands and hedgerows and other semi-shaded places, mostly in SW England and Wales, but extending north to W Scotland.
History: Named *All-heal* as the leaves have antiseptic properties and were used to cover open flesh-wounds before bandaging became common.

COMMON OR PERFORATED ST JOHN'S WORT
H. perforatum Pl. 3

An evergreen perennial growing 2 ft. tall with a two-edged erect stem. The small elliptic-oblong leaves are dotted with pellucid glands, while tiny black dots appear on the sepals and petals of the golden-yellow flowers. In bloom July–September.

Habitat: Common or shady banks, also in deciduous woodlands and thickets throughout England; less common in Scotland and Wales and in N England, usually growing in calcareous soils.

History: To this plant the French gave the name *mille-pertuis*, a thousand perforations, and with cells of caproic acid present in the flowers as well as in the foliage, the whole plant emits an unpleasant smell of wet fur when handled, though fruity undertones may sometimes be observed.

The plant takes its country name from St John the Baptist, for it is usually in bloom on the feast day of the saint (24th June). The red sap present in the stems, leaves and flowers was believed to represent the blood of the saint. At one time it was grown near the entrances to houses to deter evil spirits and it was the custom in Wales to fasten a sprig on doors on St John's Day. Of all plants, *Hypericum* has most medicinal properties. An infusion of its leaves is valuable for coughs and to prevent young children from bedwetting. Gerard said that the oil obtained from the leaves had valuable antiseptic properties and should be applied to flesh-wounds caused by 'a venomed weapon'. 'Hypericon was there – the herb of war'. Modern medical practitioners recommend its use to treat bed-sores in the form of an ointment prepared from the flowers and leaves and mixed with olive oil.

SLENDER ST JOHN'S WORT
H. pulchrum

A slender perennial growing about 12 in. tall with red stems and bearing its heart-shaped leaves in pairs. It is a hairless plant with a rounded stem, partly clasped by the leaves which are covered in transparent resinous glands like *H. androsaemum*. The flowers are golden-yellow, red on the

Slender St John's Wort

underside and with dark red dots at the petal-edges. The sepals, about one third the length of the petals, also have blackish-red dots at the edges. In bloom July–August.

Habitat: Enjoys a dry situation, a bank or mountainous slope and a sandy, non-calcareous soil. Common in East Anglia and SW England.

HAIRY ST JOHN'S WORT *H. hirsutum* Pl. 3

An erect downy herbaceous perennial growing 1–2 ft high with a round stem, covered in hairs. The slightly stalked, ovate, hairy leaves are covered in pellucid resinous glands but are without marginal glands. The narrow sepals are fringed with stalked glands. The pale yellow flowers are smaller than those of *H. perforatum* and are glandular only at the petal-tips. The stamens are fewer, there being only 7 to 9 in each bundle. The flowers stand more isolated and are visited by few insects, most of which are perhaps deterred by the resinous glands. Self-fertilisation takes place when the flower closes upon fading. In bloom July–September.

Habitat: In woodlands and hedgerows, especially in chalk or limestone soils, throughout the British Isles.

MOUNTAIN ST JOHN'S WORT *H. montanum*

A stiff, short-growing perennial herb with less downy stems than *H. hirsutum* and with broader sessile leaves, hairless on the upper surface and with black resinous glands at the margins only. The pale yellow flowers are borne in a dense inflorescence, the petals having glands at the edges like the leaves. The flowers are more pleasantly fragrant than those of other species. Glandular teeth are present on bracts and bracteoles. In bloom July–September.

Habitat: Present in lower mountainous scrubland and on cliffs, usually on calcareous soils and mostly in SW England and Wales, but extending northwards to Ayrshire.

MARSH ST JOHN'S WORT *H. elodes* Pl. 3

A low-growing stoloniferous perennial, in habit unlike any other species. The mat-forming stems root at the nodes. The leaves are oblong or heart-shaped, covered in down, likewise the sepals of the yellow flowers, which have crimson glandular teeth. The flowers are spice-scented. In bloom June–September.

Habitat: Present in ponds, bogs and low-lying ground, especially in the Lake District and NW England.

PINK AND CARNATION FAMILY   Caryophyllaceae

Annual or perennial herbs of northern temperate regions with opposite simple, entire, exstipulate leaves and stems swollen at the leaf-nodes. The inflorescence is usually terminal on the main stem. Flowers usually with 4–5 petals and sepals joined at the base, often forming a long tube or bladder. The family may be divided into two groups:

(1) Alsinoideae, which includes *Cerastium* and *Stellaria*, bearing unscented, shallow flowers with the nectar readily available to short-tongued insects.

(2) Caryophylloideae which includes *Dianthus* and *Lychnis* bearing rose- or flesh-coloured flowers, pollinated by Lepidoptera, as only the longest-tongued insects are able to reach the honey secreted at the base of the elongated tubes. The flowers have a rich clove-like perfume, similar in each genus, and in those with an extended tube, it is most pronounced in the evening, thus attracting the night-flying Lepidoptera.

SILENE   Annual or perennial herbs often viscid, with the 5-toothed calyx inflated like a bladder. The leaves opposite, exstipulate. Flowers borne solitary or in cymose inflorescence; petals 5 with narrow claws; sepals with 5 free teeth, joined below. Stamens 10; styles 3 or 5, alternating with the sepals. Fruit, a 3-chambered capsule; seeds black. The viscid property protects flowers but the plant is not insectivorous.

BLADDER CAMPION
Silene vulgaris
Syn: *S. inflata*

A usually hairless, glaucous, bushy perennial growing 2 ft tall with oblong, acute leaves, the lower of which have short stalks. The white flowers, borne in a many-sided drooping panicle, have a reddish-green bottle-shaped calyx with 5 broad teeth. The flowers, with their deeply cleft petals, open in the evening and emit a soft clove-like scent. They are visited by nocturnal Lepidoptera. In bloom July–August.

Bladder Campion

Plate 3
St. John's Wort 1 Marsh St. John's Wort 2 Hairy St. John's Wort
Family 3 Sweet Amber 4 Common St. John's Wort

Plate 4
Pink & Carnation Family (i)   1 Night-flowering Catchfly   2 White Campion 3 Nottingham Catchfly

Habitat: Common in cornfields, on waste ground and in hedgerows throughout the British Isles.

BRECKLAND CATCHFLY *S. otites*

A perennial with an erect viscid stem 12 in. tall and few, spath-like leaves covered in dense hairs; the lower leaves formed in rosette fashion. The flowers are small and pale yellowish-green and are borne in a broad spike. The female flowers are without stamens (10 in the male) but 3 styles are present. The calyx is 10-veined with 5 short teeth. The flowers are nectar-secreting and open at night when they are sweetly scented. In bloom June–July.

Habitat: Found only in the Brecklands of Norfolk, Suffolk and Cambridgeshire where it is now rare.

NOTTINGHAM CATCHFLY *S. nutans* Pl. 4

A branching perennial herb growing about 18 in. tall, which is hairy and viscid. The basal leaves are stalked; the upper lanceolate and subsessile and hairy all over. Flowers white or pale pink with cylindrical pubescent calyx, swollen in the middle. The flowers open fully in the evening when they are fragrant for three nights and are visited by night-flying Lepidoptera. The flowers have deeply cleft petals and droop in one direction. The 5 stamens ripen on the first two nights, the styles protruding on the third. In bloom June–July.

Habitat: On rocks and cliffs and dry ground throughout the British Isles, especially the Channel counties and the Midlands. Widespread in Derbyshire and Nottinghamshire; also N Wales.

History: So named because it was most common on the rocks upon which Nottingham Castle stands. Closely related to the *Lychnis* and in earlier times the leaves of both species were used as wicks.

NIGHT-FLOWERING CATCHFLY *S. noctiflora* Pl. 4

Syn: *Melandrium noctiflorum*

An annual growing 12 in. high and like *S. nutans* is pubescent and viscid with a branched flowering stem. The lower lanceolate leaves, 4 in. long, have a small stalk; the acute upper leaves are sessile. The large flowers of pinkish-white, shaded yellow on the outside, open at night or whenever the temperature falls below 47° F, and emit a delicious clove-like perfume which attracts the night-flying Lepidoptera. In bloom July–September.

Habitat: Common in cornfields where the soil is of a sandy nature. Widespread throughout Britain, but rare in Ireland.

WHITE OR EVENING CAMPION   *S. alba*   Pl. 4

Syn: *Lychnis alba; Melandrium album*

An annual or biennial herb with a woody rootstock from which arises a much-branched stem. It grows 18 in. tall with ovate-lanceolate leaves covered in soft hairs and its pure white flowers, 1 in. across, are borne in loose cymes. The calyx is dark green, inflated and is almost twice the length of that of the Red Campion which also occasionally bears a white flower. Though not completely closed by day, the flowers expand at night. This, and the fact that they are sweetly scented at night, make them more attractive to nocturnal Lepidoptera. In bloom May–June.

Habitat: Most widespread species, found in hedgerows and deciduous woodlands in all parts of the British Isles, except far N of Scotland.

History: Michael Drayton, in the *Polyolbion*, suggests several of the most sweetly scented flowers of the *Dianthus* species for making a pot-pourri:

> Sweet William, Sops-in-Wine, the Campion and to these
> Some lavender they put, with rosemary and bays,
> Sweet marjoram with her like sweet Basil, rare for smell . . .

DIANTHUS Annual or perennial herbs with narrow grass-like glaucous leaves and of matted habit. Flowers borne terminal, solitary or in clusters. Calyx tubular and 5-toothed with imbricating basal bracts. Petals 5, rose- or flesh-coloured with long claws, toothed or cut, hairy or smooth. Stamens 10; styles 2. Capsule cylindrical, opening by 4 valves. Pollination by butterflies and moths. Plants of the northern temperate regions, tolerant of extreme cold.

CHEDDAR PINK   *Dianthus caesius*   Pl. 5

Syn: *D. gratianopolitanus*

A grey-green, matted perennial with short, linear, sharply pointed leaves, rough at the edges. The rose-pink flowers are borne solitary on 4-in. stems, the bracteoles a quarter as long as the calyx, the petals daintily cut and downy with a delicious clove fragrance which will scent the air around. It is a long-lived plant and blooms with freedom over an extended period. It is visited by butterflies and also by night- and day-flying hawkmoths.

Habitat: A rare occupant of the limestone cliffs of the Cheddar Gorge, to be seen in almost inaccessible places, exposed to sunshine and wind. The late Mr Will Ingwersen included it among the dozen best alpine plants for the garden. Icomb Hybrid, raised in the Cotswolds by Capt. Simpson-Hayward, is an improved form. The late Montagu Allwood crossed *D. caesius* with the *Allwoodii* pinks to begin a new race of Alpine Pinks known as *D. allwoodii alpinus* which have scented flowers.

COMMON PINK *D. plumarius* Pl. 5

A tufted perennial, glabrous and glaucous with grass-like leaves, rough at the edges. The flowers, pale pink or white and about 1 in. across, are borne solitary or in cymes of 2–5. The bracteoles are a quarter as long as the calyx-tube which has narrow teeth. The petals are downy and cut as far as the middle to present a feathered effect, hence its botanical name. Pollination is by butterflies by day, hawkmoths at night. In bloom July.

Habitat: Mostly on old walls, especially of monastic foundations and ancient castles, the plants enjoying an open, sunny situation and with their roots in limestone rubble or mortar.

History: It is generally conceded that the modern Pink is descended from *D. plumarius* and the Clove-scented Pink and the Carnation from *D. caryophyllus*, both of which are believed to have been introduced into England after the Norman invasion of 1066. To this day plants are to be found naturalised on the walls of Norman castles as at Rochester, Pevensey and Dover, and on the walls of ancient monastic buildings.

The flower is portrayed in a picture by Hans Holbein (the younger) which hangs in the Dahlem Museum, Berlin. It is a delightful painting of George Gisze, a Hanseatic merchant of Danzig, a man who was much interested in flowers and in cut-glass and Oriental rugs. Holbein shows him with a cut-glass vase containing three Pinks, and has enhanced their bright pink colouring by painting the merchant's sleeve exactly the same shade, a colour which he used in several of his paintings. It was executed in London in 1532 and from it we can understand the great esteem in which the flower was held at the time, for in the picture it is prominently displayed. In the same gallery hangs Jan van Eyck's painting of *A Man with a Pink* which was executed almost a century earlier, and when comparing the two it is not difficult to see how much the flower had improved both in form and colour during the time that had elapsed between the execution of the two paintings.

In medieval art, the Pink was usually the symbol of divine love and signified that a lady was engaged to be married. In the painting *Lady with a Pink* by the German-born Hans Memling, dated to the mid-fifteenth century and now in the Metropolitan Museum of Art, New York, the girl is shown holding a Pink in her left hand. It is a single bloom and from its rose-pink colour and fringed petals, it would appear to be a bloom of *D. plumarius*.

Rembrandt's *Lady with a Pink* most probably portrays his son Titus' future wife. It hangs in the same Museum of Art and was painted about 1668. The artist has depicted the light radiating from the flower.

To the poets the flower was the emblem of perfection, as Robert Burns suggested in his poem to Mary, which he called 'The Posie':

... I will pu' the Pink, the emblem o' my dear;
For she is the pink o' womankind, and blooms without a peer ...

D. plumarius is accurately shown together with the perennial Pea and
the Daisy in an illuminated page from a *Book of Hours* by the Master
of Mary of Burgundy. It dates from about 1480 and is now in the
Bodleian Library at Oxford.

CLOVE-SCENTED PINK, CARNATION
D. caryophyllus   Pl. 5

A glabrous much-branched perennial, distinguished from *D. plumarius*
by the smooth edges of its leaves and larger flowers which measure
1½ in. across. They are borne mostly singly on 12-in. stems and are
flesh-pink or rose with the petals glabrous and deeply notched. They
are visited during the day by butterflies, and at night by hawkmoths.
These insects are attracted by the powerful clove perfume produced by
the eugenol present which, as Darwin pointed out, is especially pronounced at night. In bloom July–August.

Habitat: Mostly on old walls, especially of monastic foundations and
ancient castles, the plants enjoying an open, sunny situation and with
their roots in limestone rubble or mortar.

History: The Athenians held the plant in such high esteem that they
named it *Di-anthos*, Flower of Jove, awarding it the highest honour. It
was the chief flower for the making of garlands and coronets, hence its
early English name of 'Coronation' from which the name Carnation is
derived.

It was the opinion of Canon Ellacombe, the Victorian authority on
plants and their history, that *D. caryophyllus*, (and *D. plumarius*)
native to S Europe, reached England with the Norman invasion,
possibly attached to stones imported from northern France (Caen) for
the erection of castles and houses of worship. Canon Ellacombe
reported having seen *D. caryophyllus* in bloom in 1874, on the walls of
the Conqueror's Castle of Falaise (where William was born) and in
England it can be seen to this day growing on the walls of the Norman
castles of Dover and Rochester in Kent, and on the walls of Fountains
Abbey in Yorkshire where it blooms in July. Indeed, several of the
writers of old believed its country name of *Gillyflower* to be derived
from *July-flower*. Herrick was of this opinion:

> A lovely July flower
> That one rude wind or ruffling shower
> Will force from hence and in an hour.

Sir Francis Bacon also wrote that 'in July come gillyflowers of all
varieties . . .' from which it would appear that the reference was to

Plate 5
Pink & Carnation Family (ii)

1 Cheddar Pink
3 Common Pink

2 Clove-scented Pink

Plate 6
Geranium Family   1 Common Storksbill   2 Musk Storksbill
3 Herb Robert

Double Pinks or Carnations (which Gerard called the Great Double Carnation) rather than to Single Pinks, for he suggests growing the Pink for earlier flowering. In his essay *Of Gardens* he wrote that in May and June 'come pinks of all sorts, especially the blush pink'.

With the scent of the flowers of *D. caryophyllus* resembling the perfume of the clove, the plant was known to the French as *giroflier*, a name which accompanied the plant to England with the Normans. Chaucer, in the *Prologue to the Canterbury Tales* (1386), referred to the clove-scented flowers which were at that time much in demand for flavouring wine and ale (p. 20) possibly a French custom, and Pinks, known as *Soppes-in-wine*, were to be found until the end of the sixteenth century, growing in tavern gardens. Chaucer mentions this fact in *The Tale of Sir Topas*:

> Ther springen herbes grete and smale,
> The licoris and the setewale;
> And many a clove gilofre,
> And notemuge, to put in ale,
> Whether it be moiste or stale . . .

Chaucer's spelling of the word differs but little from the early French, while the poet Shelton wrote of the *Ielofer amyable*. By Shakespeare's time, it had become *Gillovor* or *Gillyflower*, the word being used for all clove-scented flowers, e.g. Queen's Gillyflower (*Hesperis matronalis*) and Stock Gillyflower.

Chaucer's reference to nutmeg is interesting in that it may refer to the old Nutmeg Clove Carnation in cultivation at the time and bearing a small semi-double flower with a distinct smell of nutmeg. It was later known as *Fenbow's Nutmeg*.

By Tudor times, there were two distinct groups of *Dianthus*, those with single flowers known as the Pinks and descended chiefly from *D. plumarius*, and those bearing double (or semi-double) flowers, offspring of *D. caryophyllus*, known as *Gilloflowers*. Both were scented, but the offspring of *D. plumarius* were without the clove perfume and had more deeply cut petals.

In the *New Herbal* of 1578, Lyte distinguished between the two forms by his use of the word 'Coronations' and of 'the small feathered Gillofers, known as Pynkes, Soppes-in-Wine and small Honesties'. Writing at the same time, Spenser separates the two in his lines from *The Shepherd's Calendar* (p. 20). Spenser also differentiates here between the Pink, the Carnation and Sops-in-wine which may be a distinctive form of the *Dianthus* (possibly a small semi-double Pink), for it is doubtful whether the earlier reference to 'the Pinke and Purple Cullambine' really meant Columbines of pink and purple colouring, as the work 'pink' signifying the particular colour was not introduced into the English language until late in the eighteenth century. Before that, the colour we now call *pink* was always described as *flesh* or *blush* or even *carnation*, as in Byron's line:

Carnation'd like a sleeping infant's cheek.

The earliest Carnations (or semi-double Pinks) bore flowers of the colour of flesh-wounds, which may be described as deep 'pink' and are alluded to in *Henry V* when, in Mistress Quickly's house in Eastcheap at the time of Falstaff's death, the Boy says, '. . . and (he) said they were devils incarnate', to which Mistress Quickly replies: ' 'A could never abide carnation; 'twas a colour he (Falstaff) never liked'. The name is derived from the Latin *carnis*, flesh.

To the Elizabethan poet Shenstone, both the colour and the perfume of the Carnation was most agreeable:

Let yon admired carnation own,
Not all was meant for raiment, or for food
Not all for needful use alone;
There, while the seeds of future blossoms dwell,
'Tis colour'd for the sight, perfumed to please the smell.

The Roman historian Pliny described how the Clove-scented Pink was discovered in Spain during the reign of the Emperor Augustus, when it received considerable attention in Rome as a flavouring for wine. The plant had been discovered in the part of Spain bordering the Bay of Biscay, at that time inhabited by the warlike Cantabri after whom the plant was originally named. As late as the mid-sixteenth century, Dr William Turner, a close friend of Bishops Latimer and Ridley and himself Dean of Wells, justifiably called the Pink *Cantabrica gelouer* in his *New Herbal*, the first part of which was published in 1551 and dedicated to Queen Elizabeth.

In his *Herbal* of 1597, Gerard distinguished between the *Carnation Gilloflower* (the Great Double Carnation) and the *Clove (or Pink) Gilloflower* (the name *Dianthus* was not in use until Linnaeus had compiled his binomial system of plant classification), from which it would appear that by then the Carnations had not the same perfume as the Pinks, though of the Great Double Carnation, Gerard said its flowers 'had an excellent sweet smell' though perhaps not of clove.

That there were many different species and varieties is confirmed by Gerard who said: 'A great and large volume would not suffice to write of every one at large . . . and every year every climate and every country bringeth forth new sorts . . .'

Gerard also singled out the 'wild (native) Gilloflowers' as they had smaller flowers than the 'Carnations and Clove Gilloflowers'. Earlier, Tusser had written of 'pinks of all sorts' and at a later date, Milton included the 'white pink' – possibly the Old Fringed, which is still obtainable – among other scented flowers to be strewed over the hearse of Lycidas.

That there were 'streaked gillyflowers' in Elizabethan times is confirmed by Shakespeare, and in the *New Book of Flowers* by Maria Merian, a copy of which is in the British Museum, there is a colour illustration which shows the 'streaked gillyflowers' much as we know the Flaked Carnations of today.

In *A Winter's Tale* Perdita speaks of:

> The fairest flowers o' the season
> are our Carnations and streak'd Gillyflowers.

Later in the play, Polixenes exhorts Perdita to 'make her garden rich in gillyvors'; Shakespeare may have brought the plants with him from Gerard's garden and those of the famous in London, to plant in his own garden at Stratford, and these may have included all those Pinks of clove scent (the Gillyflowers and possibly the double form of our native Cheddar Pink) as well as a number of choice Carnations including the Old Crimson Clove. Shakespeare clearly distinguished one from the other.

At the time of Shakespeare's death, Carnations and Pinks were counted among the most popular of all garden plants. Gerard said that '. . . they are well known to most, if not to all' and William Lawson in *The Country Housewife's Garden* (1618) said 'I may well call them the king of flowers, except the rose'.

They were popular, perhaps for their hardiness, and though native to S and E Europe, they quickly became acclimatised to English gardens where they were most suitable for the small 'knot beds' of the time.

The poet John Fletcher commented on their hardiness:

> Hide, O hide, those hills of snow
> Which thy frozen bosom bears,
> On whose tops the pinks that grow
> Are of those that April weares.

By the beginning of the seventeenth century, the Pink and Carnation had reached their peak of popularity. Parkinson, writing shortly after Shakespeare's death, said, 'What shall I say to the Queene of delight and of flowers, carnations and gillyflowers, whose bravery (hardiness), variety and sweete smell joyned together, tyeth every ones affection with equal earnestness both to like and to have them?' Writing about the same time, Gervase Markham in *The English Husbandman* (1613) said, 'Gilliflowers are of all other flowers, most sweet and delicate' and in *The Complete Gardener's Practice* (1664), Stephen Blake wrote, 'Carnation gilliflowers for beauty and delicate smell and excellent properties, deserve letters of gold. I wonder that Solomon did not write of this flower when he compared his spouse to the Lily of the Valley'.

Exactly when the *Dianthus* received its name *Pink* is not known. Mr Stanley B. Whitehead believes the name to be a derivation of the Celtic *pic* meaning peak, possibly the 'peak of perfection' among flowers; on the other hand, Mr L. J. Brimble in his *Flowers of Britain* suggests that the name was obtained from the verb *to pink* or *to pierce*, indicating the extremely serrated petals of *D. plumarius* and its off-spring, the Old Fringed Pink and Mrs Sinkins. Again, it has been suggested that the name comes from *Pinksten*, German for Pente-cost, for it is at this time (late Whitsuntide) that the Pink comes into bloom, and on the Continent the plant is known as the *Whitsun Gilly-flower*.

The native pinks of the British Isles remained very much neglected until Parkinson (1629) said that '. . . some grow upright like gilliflowers or spreading over the ground'. Then, during the reign of Charles I, a number of Pinks were introduced from France by Queen Henrietta Maria, to whom Parkinson dedicated the *Paradisus*, for the Pink is said to have been her favourite flower. But by the time of the Restoration, John Rea in his *Flora* remarked that 'there are few carnations to be found in any of our gardens'. However, in a later edition of his work he listed 360 varieties of Carnation and Pink and added:

> But yet if 'ask and have' were in my power,
> Next to the Rose, give me the Gillyflower.

Many of the old Pinks still survive. One with the name *Sops-in-wine* is known to have existed in the fourteenth century and may be that mentioned by Chaucer. It is believed to have reached England from a monastery near Orleans in France and may have come with the Nor-mans or shortly afterwards. It is to be found in cottage gardens in Berkshire and has fringed petals and a jet-black centre. A Pink of the late fifteenth century is Caesar's Mantle, the Bloody Pink of early Tudor gardens, bearing flowers of deepest crimson covered in a black grape-like 'bloom'. Of a similar period is Unique, a Pink of Painted Lady type bearing single flowers of red ground colouring, flecked all over with black and pink. Mrs Sinkins, discovered in the garden of the Workhouse at Slough and named after the wife of the custodian, still survives and has the true clove perfume. Another is Earl of Essex, a Pink of ancient origin, being a very double form of *D. plumarius* and of similar colouring.

Gerard said that a conserve made of the flowers (of *D. caryophyllus*) with sugar 'is exceeding cordial and wonderfully above measure doth comfort the heart . . .'

HERNIARIA Small prostrate perennial herbs with narrow, opposite leaves. Flowers minute, borne in axillary clusters. Sepals and petals 5, latter resembling barren filaments and shorter than the blunt sepals. Styles 2-branched with 2 stigmas. Flowers self-pollinated and scentless; leaves with a pleasant sweet smell.

SMOOTH RUPTURE-WORT   *Herniaria glabra*
An annual or biennial forming a mat of small oval glabrous leaves. The flowers are borne in axillary clusters of up to 12, with blunt hairless sepals and minute white petals. Self-pollinated. The leaves have a sweet refreshing scent resembling fresh hay and contain coumarin. In bloom July–September.
Habitat: A rare native, found in a few small colonies in cropped turf where the soil is of a sandy nature, in parts of Suffolk and Cambridgeshire, extending into Lincolnshire.
History: It takes its botanical name from the belief that an infusion of the leaves acted as a cure for hernia or rupture.

GOOSEFOOT FAMILY   Chenopodiaceae
Common weeds, annual or perennial herbs, distributed about the temperate regions of the world, mostly by the seashore or in salt marshes. They have exstipulate leaves, lobed or toothed, and usually grooved stems. Included in the family are the Beetroot, Spinach and Sugar Beet, valuable commercial plants. Flowers mostly hermaphrodite, borne in cymes on a branched inflorescence. Perianth 3-5 lobed, imbricate and persistent. Mostly wind-pollinated; several self-pollinating, scentless.

CHENOPODIUM (Goosefoot) Mostly herbs. Stems angular, 6 in.–2 ft tall, often mealy, occasionally striped red or white. Leaves alternate, thick and fleshy, flat, entire or lobed. Flowers minute, borne in axillary panicles, perianth 3-5 cleft; stamens and stigmas 3–5. Leaves of several species unpleasant-smelling.

STINKING GOOSEFOOT   *Chenopodium vulvaria*
A low-growing grey-green annual with fleshy ovate leaves which are usually notched near the base; sometimes entire. The underside is covered in greasy meal which contains trimethylamine, present in Hawthorn blossom (though lightened with traces of anisic acid), which gives off the obnoxious smell of decayed fish when handled and remains on the hands after washing. The plants also give off 'free' ammonia. The yellowish flowers are borne in dense clusters at the base of the leaves. In bloom July–August.
Habitat: By the margins of the seashore or salt-marshes throughout England, but especially Kent and Essex.

LIME FAMILY   Tiliaceae

Trees or shrubs mostly of tropical and temperate regions of the world
with alternate stipulate leaves showing 2-ranked arrangement, entire,
toothed, rarely lobed. Flowers regular, borne in terminal cymes.
Sepals 5, free or united; petals 5, stamens 10 or more with filaments free
or united at base; style 1. Anthers with 4 pollen sacs which distinguishes
the family from Malvaceae. Fruit, a capsule, drupe or nut.

TILIA (Lime or Linden Tree) Trees with stellate hairs. Leaves
alternate, stalked, cordate, serrate. Flowers yellowish-green, borne in
terminal cymes on the young growths with leafy bract attached to the
stalk. Sepals and petals 5; stamens numerous, free or in bundles. Style
with 5-lobed stigma. The powerful scent of the flowers, due to the
presence of the lily-scented fornesol, attracts all types of insects, and
those with short probosces can readily obtain the honey secreted and
lodged in the hollow sepals. Self-fertilisation is prevented by the
stamens bending outwards and not coming into contact with the
stigma.

From the flowers a medicinal tincture is made with spirits of wine
and used to relieve headaches and giddiness. The flowers, which
contain a fragrant volatile oil soluble in alcohol, may be used for
making wine and a fragrant distilled water may be obtained from them.
From the flowers, Gilbert White made a nourishing 'tea', while the
bark yields a soft mucilage, valuable for healing skin burns.

SMALL-LEAVED LIME   *Tilia cordata*

A large tree with handsome, smooth bark and twigs. The finely toothed
cordate leaves, thick and leathery, have tufts of woolly hairs in the
veins on the underside. The yellowish-white flowers are borne in
spreading cymes with the stamens as long as the petals. The flowers are
visited by all forms of insect life. In bloom June–July.
Habitat: Present in all parts of England and Wales but not naturally in
Scotland. The tree shows a preference for limestone regions, hence its
common name. It is especially prolific in parts of Derbyshire and
Staffordshire; and in Devon and Dorset.

COMMON OR EUROPEAN LIME   *T. europaea*
Syn: *T. vulgaris*

A large tree, indigenous to England and to most parts of Europe. It has
smooth dark brown bark and the trunk is often covered in bosses near
the base. The leaves are ovate or cordate, hairless above but with tufts
of hairs in the axils of the nerves or veins on the underside. The flowers,

borne in pendulous cymes, are yellowish-white with the stamens equalling the number of petals, and are powerfully scented. In bloom June–July.

Habitat: Though it has a liking for a limestone soil, it is to be found in most parts of the British Isles, except N Scotland. It is common in deciduous woodlands and copses and has been widely planted in parklands and gardens.

History: 'The lime at dewy eve diffusing odours', wrote William Cowper. Of the tree Dr Turner wrote (1560), 'It groweth very plenteously in Essekes, in a park within two miles from Colichester'. In his *Herbal*, Gerard (1597) said that 'It groweth in some woods in Northampton-shire; also near Colchester . . . and in My Lord Treasurer's Garden in the Strand'. Though a native tree, *Tilia vulgaris* must have been rare, for Evelyn complained of having to import the young trees from Holland, saying 'It is a shameful negligence that we are not better provided of nurseries of a tree so choice and universally accepted'.

Pliny said that in his day chaplets were made from the bark of the Linden tree and that Roman cooks used the inner bark to boil with meat which had become too salted.

In Britain, the wood was widely used for decorative carving, trees being planted on the great estates of Chatsworth and at Petworth solely for this purpose. At Chatsworth, Samuel Watson worked with superb artistry from the soft wood of the Lime, while at Petworth, Windsor and at Lyme Park in Cheshire (so-called because of the large Lime plantations there) Grinling Gibbons achieved long lasting fame for his exquisite work. Lime wood was chosen for several reasons for, besides its softness, it has a close grain and is of delicate colouring, and in addition is not troubled by worm. The wood also lasts long in water without decaying.

The Lime is an excellent tree for town planting and, at Evelyn's recommendation three centuries ago, was widely used for planting in St James's Park and in open spaces in London, together with the Sycamore and Plane trees. William Kent and Capability Brown made use of it in planning the landscapes of the great houses, for no tree is more attractive when the pale green leaves begin to unfold towards the end of April, their unique shade of green giving the colour lime-green to fashion. Then in autumn the leaves become a glorious golden-yellow.

The *Line*, or *Linden* tree as it is still called in Holland and Germany (*Unterden Linden*) is one of the most suitable of all trees for clipping into the shape of a pleasing 'walk'.

In Elizabethan times, the Lime was widely planted as a pleached alley, and it was beneath such an alley that Antonio saw Don Pedro and Claudio walking together in *Much Ado about Nothing*. The word *pleach* is taken from the French *plesser*, to plait or interweave. 'Pleach't armes', wrote Shakespeare in *Antony and Cleopatra*, and he uses the same word to describe the braiding of hair.

The word *line* is derived from the Anglo-Saxon *lind*, a shield, for as it is so easy to work, the wood was used for making shields. The Saxon *lind* became *lynde*, the word used by Chaucer and then *line*, used by Shakespeare. 'As light as leaf on lynde', wrote Chaucer in *The Clerk's Tale*; and in *The Tempest*, Prospero asks Ariel, 'How fares the King and his mistress?':

> Ariel: Confin'd together
> Just as you left them; all prisoners sir
> In the line grove which weather-fends your cell.

Here Shakespeare had in mind the dense covering afforded by an alley of pleached Limes. These trees, when pleached, grew no more than 6–8 ft tall and country housewives would hang their clothes on them to freshen and their laundry to dry. Gerard quotes Theophrastus as saying that Lime leaves were sweet to smell and were eaten by cattle.

In *The Tempest*, Shakespeare makes play of the word when Prospero tells Ariel, who enters loaded with apparel:

> Come, hang them on this line.

The perfume and the beauty of the Lime has never been expressed more gracefully than by Abraham Cowley, born when Shakespeare was but two years dead. He wrote:

> 'Mongst all the nymphs and Hamadryades,
> There's none so fair, and so adorn'd as this.
> All soft her body, innocent and white,
> In her green flowing hair she takes delight;
> Proud of her perfum'd blossoms far she spreads
> Her lovely, charming, odoriferous shades.

No tree better hides its flowers than the Lime for they are almost the colour of the leaves, a fact which Sir Aubrey de Vere pointed out in his lines 'To the Lime Flower':

> We see you not; but, we scarce know why,
> We are glad when the air you have breathed goes by.
> Hail blossoms green 'mid the limes unseen,
> That charm the bees to your honey'd screen
> We see you not . . .

MALLOW FAMILY Malvaceae

Downy herbs, trees or shrubs mostly native of the tropics and temperate regions with alternate, palmately lobed stipulate leaves. Flowers borne solitary and axillary or in cymose inflorescences. Sepals 5, united at base; petals 5, twisted in bud. Stamens numerous, in a bunch; filaments more or less united; fruits shaped like round cheeses, hence their name *cheesecakes*. Family includes *Abutilon*, pollinated by humming-birds; *Lavatera* and *Malva*, mostly self-pollinating. Few have scented attractions.

MALVA (Mallow)   Smooth or hairy annual or perennial herbs with leaves angled, lobed or dissected. Flowers borne solitary or in clusters; double calyx of 5 sepals, united at the base; petals 5, deeply notched, purple, pink or white. Stamen-tube divided at apex into numerous filaments. In several species, anthers and stigma ripen together, thus ensuring self-fertilisation. One species has scented foliage.

MUSK MALLOW *Malva moschata*
A hairy perennial growing 2–3 ft tall, the basal leaves long-stalked and divided into 5 fern-like segments. Stems and leaves covered in hairs which emit a musk-like smell when handled, a smell similar to that released by the tomentum of the Musk Rose, the Musk Storksbill and the cottage garden Musk before it lost its hairs. Flowers usually borne in graceful spikes in the axils of the leaves; they are of mallow-pink colouring and about 2 in. across. The flowers are followed by the familiar ring of seeds (like a cheese). In bloom July–August.
Habitat: Common in hedgerows and on dry banks throughout the British Isles.
History: Named by Linnaeus from a Greek word meaning soft, a reference to the hairy characteristics of stem and leaf.

Musk Mallow

GERANIUM FAMILY   Geraniaceae

Herbs or sub-shrubs present in most parts of the temperate world. Leaves opposite or alternate, toothed or lobed, usually stipulate, hairy. Flowers regular, red, pink or purple. Petals 5; sepals 5. Stamens usually 10; 5 imperfect. Ovary 3–5-chambered with 1–2 or more seeds in each chamber.

GERANIUM (Cranesbill)   Usually herbs with opposite or alternate, toothed or palmately lobed leaves. 1–2 flowered; sepals 5, alternate with the 5 petals; glands 5, alternate with the petals which are usually red, blue or purple. Carpels 5, curling upwards when ripe around a central column, shaped like a heron's, or crane's, bill. The fact that the flowers are mostly blue or purple with darker centre lines and turn towards the sun to make them more visible from a distance, means they are constantly visited by Hymenoptera; they are not scented. The leaves emit a resinous smell in *G. robertianum*. In bloom June–August.

HERB ROBERT *Geranium robertianum* Pl. 6

It has dainty frond-like leaves divided into 3–5 pinnately lobed segments and red wiry stems. It is perennial, growing 9–15 in. tall, and is entirely covered with small hairs which release a foxy or musky smell when handled. Miss Sinclair Rohde, however, considered it the most pleasantly scented of all the native members of the family, being resinous and refreshing. The pinkish-red flowers, borne in succession throughout the summer, measure less than 1 in. across. In bloom May–end of summer.

Habitat: It is found in hedgerows and about deciduous woodlands; also on old walls throughout the British Isles and to an altitude of at least 2000 ft. It is present in most parts of the temperate world.

History: The plant takes its botanical name from the Greek *geranos*, a crane, denoting the manner in which the carpels are arranged around the central column, like a crane's bill. It takes its country name from a Frenchman, the Abbé Robert, who founded the Cistercian Order in the eleventh century. From earliest times the leaves have been appreciated for their healing powers and the plant is described in glowing terms in *The Feast of Gardening*, a treatise by Master John, Gardener.

ERODIUM (Storksbill) Differs from *Geranium* in the spirally twisted awns which when ripe may shoot off to a distance from the plant. They uncurl when in contact with moisture in the ground, and then eject the seed. Annual or perennial herbs with leaves pinnately lobed. The flowers are borne solitary, in pairs, or in small clusters. Petals 5, mostly pink or rosy-purple; sepals 5, imbricate. Stamens 5, alternating with 5 scale-like staminoids.

MUSK STORKSBILL *Erodium moschatum* Pl. 6

An annual herb growing about 12 in. tall and branching at the base. The dark green 1-pinnate leaves, also the stems and footstalks, are covered in silky hairs which are often sticky when handled and release a strong smell of musk. The flowers are mauve-pink with the petals longer than the calyx, and are borne in small umbels. In bloom May–September.

Habitat: It is common on waste ground close to the sea but mostly in

S England and Ireland; also on the Yorkshire coast and in the Channel Isles.

HOLLY FAMILY Aquifoliaceae

Evergreen trees or shrubs, mostly native of the tropics and temperate regions of the world. Leaves alternate, simple, exstipulate, elliptic and often spiny. Flowers small, white or greenish-yellow, borne in axillary cymes. Sepals 3–6, united; petals 4–6; stamens equalling petals in number and alternate with them. Anthers 2-chambered; ovary superior; stigma lobed. Fruit, a fleshy drupe.

ILEX (Holly) Evergreen trees or shrubs with alternate shining, toothed or spiny leaves. Flowers white, borne axillary. Calyx small, 4–5 cleft. Petals united below. Stamens 4–5, adherent to corolla tube; stigmas 4–5. Fruit, a round red or yellow berry.

HOLLY
Ilex aquifolium
A small evergreen tree with smooth grey bark and dark green glossy leaves with waved spinous margins, those towards the top of the tree often with only an apex spine. Flowers white, 4-petalled, borne in axillary clusters, male and female on different trees. They are sweetly scented and are visited by most forms of insect life which seek the readily available nectar secreted at the base of the petals. In bloom May–August.

Habitat: Widespread throughout the British Isles; is found in deciduous wood-lands and hedgerows, also about low mountainous rocks, usually in partial shade. It requires a sandy soil and is most prevalent in Shropshire, Herefordshire and Worcestershire.

Holly

History: The tree takes its name from the Latin *aquifolium*, sharp-leaved, while the name Holly is perhaps derived from *holy*, the name used for the plant by Turner in his *New Herbal,* possibly because it was in demand for decorations at Christmastide, as it retains its leaves and berries throughout winter. Stowe, in his *Survey of London* (1598), remarked that churches, houses, and market crosses were bedecked with Holly branches at this time. It was also used for enclosing land. Columella, in his *10th Book,* recommends its planting, to the Romans, for this purpose, and the tree may have been introduced into Britain during the Roman occupation:

> And let such grounds with walls or prickly hedge,
> Thick set, surrounded be and well secured.

Evelyn has told us that his garden at Saye's Court was surrounded with a Holly hedge 400 ft long and 9 ft tall and added, 'it mocks the rudest assaults of the weather, beasts or hedge-breakers'. Evelyn also mentioned that he raised the hedge 4 ft high from the seedling stage in two years.

As many as 60 varieties are known, some with golden leaves and others with leaves edged with silver. The wood has a fine grain and will take a hard polish. It is used in cabinet-making and for engraving.

BOX FAMILY　Buxaceae

Evergreen trees or shrubs with alternate or opposite, simple, entire, exstipulate leaves. Flowers monoecious, small, borne in axillary clusters. Calyx of 4 sepals; petals none. Stamens 4, opposite sepals. Ovary superior; styles free. Fruit, a drupe or capsule.

BUXUS (Box) Evergreen trees or shrubs. Leaves opposite, exstipulate. Flowers monoecious, axillary, bracteate, borne in clusters with a single terminal female and several males. Males with 4 sepals; 4 stamens. Female with 3-celled ovary; styles short. Fruit a capsule.

BOX TREE　*Buxus sempervirens*
A small slow-growing evergreen tree or shrub with grey bark and twigs covered in pubescence. The oblong, shining pale green leaves are about 1 in. long and are shortly stalked. When brushed against, they emit a musky smell which is most pronounced when the weather is dry and warm. The tiny greenish-white flowers are borne in crowded clusters and are scentless. In bloom March–May.
Habitat: It is found throughout England and in Ireland but is mostly confined to a few localities where the soil is over chalk or limestone, as at Boxted (now Buxted) in Sussex, Box in Wiltshire, Boxgrove in Worcestershire; also in N Devon.
History: The plant was introduced into English gardens during

Elizabethan times to surround small beds of flowers and herbs, for not only was it slow-growing with a neat compact habit but was evergreen and withstood hard clipping. It provided the plants with valuable protection against the cold winds of winter. For edging, Parkinson recommended it 'above all other herbs' for it did not straggle over the paths and beds as did Rosemary and Marjoram, while 'it suffered not from frosts in winter nor drought in summer'. The most compact form is the Dutch or French Box, the subspecies *B. suffruticosa*. Leonard Meager in *The English Gardener* (1688) recommended it for 'knots' as 'it is the handsomest, the most durable, and the cheapest' of plants.

Of the Common Box, Evelyn wrote in his *Diary* for 24th July, 1655, 'I went to Boxhill (in Surrey) to see those rare natural bowers, cabinets and shady walks in the box copses' and in *Twelfth Night*, Shakespeare recalled the natural Box bowers which were to be found about the Warwickshire and Worcestershire countryside (as at Boxgrove) when, in Olivia's garden Maria the maid said, 'get ye all three into the Box tree'.

Pliny mentioned that in his garden in Tuscany, Box was clipped into the shape of animals, and during the reign of William and Mary, topiary was carried to such extremes that in the *Spectator* of 25th June, 1712, Addison launched his celebrated attack. 'Our trees arise in cones, globes and pyramids', he wrote, 'we see the marks of the scissors upon every plant and bush'.

Until the end of the nineteenth century, the Box was associated with funerals. Chaucer in *The Knight's Tale* wrote:

> The Box tree or the Aschen dead or cold

while Wordsworth described the quaint North Country custom of placing a bowl of Box sprigs for the mourners, who would take one and throw it into the grave:

> Fresh sprigs of green box-wood, not six months before,
> Filled the funeral basin at Timothy's door.

PEA FAMILY   Leguminosae (Papilionaceae)

The third largest order of flowering plants present throughout the world but mostly confined to the warmer parts. Trees, shrubs, climbers, water-plants, the roots having tubercles which enable the plants to store additional nitrogen. Stem usually erect, though some climbing. Leaves usually alternate, often pinnate or 3-foliolate. Flowers borne in racemes, panicle or spike; papilionate, with standard petal, 2 lateral petals (wings) and 2 lower petals (keel) connected, which protect the reproductive organs in wet weather. The flowers are mostly pale yellow or pale pink and where scented are visited by butterflies, though all are visited by Hymenoptera. The soft vanilla-like perfume with lemony undertones, is similar in all the Legumes which bear scented flowers, e.g. *Lathyrus* (Sweet Pea); Laburnum; *Vicia* (Bean); *Ulex* (Gorse). In those flowers where the stamens and stigma emerge from the carina and return within it, e.g. *Trifolium*, there are repeated visits by bees, while with those flowers whose organs are confined under tension and explode, a single visit by the bee is more usual. Fruit, a dry or fleshy, curved or straight pod; seeds often large as in Lupin, Laburnum.

LUPINUS (Lupin) Annual or perennial herbs or shrubs mainly found in North America and the Mediterranean; leaves simple or 3- to many-foliolate; stipules adnate to base of leaf-stalk. Flowers borne in terminal racemes; calyx 2-lobed; standard round; wings sickle-shaped; keel ending in curved beak. Stamens 10, united in one bundle. Pod, silky-haired.

TREE LUPIN *Lupinus arboreus* Pl. 7
Introduced from Pacific USA mid-eighteenth century. Forms a large bush 3 ft tall and of same width, with attractive bright green palmate leaves, silky on the underside and divided into 7–11 lanceolate-linear leaflets. Flowers yellow, borne in large erect spikes with a scent similar to that of Laburnum and Sweet Pea. Pods 8–12 seeded, hairy. In bloom June–July.
Habitat: Naturalised on waste ground of a sandy nature throughout the British Isles but mostly S England.

ULEX (Gorse) Evergreen shrubs with leaves small, often reduced to spines. Flowers golden-yellow, borne axillary, with small bracts. Calyx 2-lipped; lower lip 3-toothed. Stamens 10, alternately short with versatile anthers. Pods hairy and when ripe they explode with a loud pop.

GORSE, FURZE *Ulex europaeus* Pl. 7
A glaucous, spiny shrub growing 8–9 ft tall and of same width. The dark green stems are furrowed; its leaves 1-foliolate and hairy. The flowers have wings longer than the keel; the calyx has spreading black hairs. The scent is refreshing, like vanilla with undertones of pineapple, also described as like fresh coconut or orange. In bloom throughout the year, especially March–August.
Habitat: Throughout Britain in rough grass, on dry banks and on heathlands, usually in poor soils of an acid nature. There is a valuable garden form, *flore dlena*, of great beauty.

Gorse

History: Of Common Gorse, William Howitt wrote:

> The crackling of the Gorse flowers near,
> Pouring an orange-scented tide
> Of fragrance o'er the dessert wide.

The late Robert Bridges, Poet Laureate, also found its fragrance equally delicious in his lines from 'The Hill Pines are Sighing':

> But deep in the glen's bosom
> Summer slept in the fire
> Of the odorous gorse-blossom
> And the hot scent of the brier.

It is said that Linnaeus, who has achieved everlasting fame for his classification of plants, upon visiting England in 1736, fell on his knees to thank the Creator for the sight of Putney Heath ablaze with Gorse blossom.

CHAT WIN   *U. gallii*

A smaller version of the Common Gorse, rarely growing more than 2 ft tall and making a rounded plant of similar width, the drooping branches armed with spines 1 in. long. The flowers have smaller bracts and are smaller in all respects than those of *U. europaeus* while their scent is not so pronounced, being soft and sweet. It flowers late. In bloom July–November.

Habitat: Widespread throughout England and Wales especially in the west; but in Scotland, only in the south. Usually on moorland and lower mountainous slopes.

ONONIS (Rest-harrow)   Small perennial shrubs growing 12 in. or less in height with pinnate 3-foliolate leaves, which, like the stems, are hairy and viscid. The flowers pink, without nectar. Calyx 5-cleft; standard broad; keel incurved and pointed. Pod usually 1–4 seeded, shorter than calyx.

CREEPING REST-HARROW
Ononis repens

A prostrate woody perennial, hairy and viscid, usually without spines. The leaves are small and oval (or trefoil) and are covered in glandular hairs which release a resinous smell when bruised.

Creeping Rest-harrow

The flowers are pink and unscented, with the keel and wings of similar length. In bloom July–September.

Habitat: Common on dry hilly pastures especially over chalk or limestone as in Sussex, Dorset and the Cotswolds. Also present by the seashore.

History: It takes its name from the Greek *onos*, an ass, for its matted, woody stems brought the ox- or ass-drawn plough to a standstill. Common in the Mediterranean countries, it may have reached Britain with the Roman invasion and is an obnoxious weed, especially to the farmer of old who used a horse-drawn plough or harrow.

ERECT REST-HARROW   *O. spinosa*

Unlike the almost prostrate *O. repens*, it grows 18 in. tall and is a woody perennial, less viscid and more resinous-scented. It is also spiny while the stems have a double line of hairs. The flowers, which are unscented, are reddish-pink, with the wings shorter than the keel. In bloom all summer.

Habitat: Waste ground in England and Wales (rare in Scotland and Ireland), especially where the soil is heavy.

TRIGONELLA (Fenugreek) Annual or perennial herbs with pinnate leaves divided into 3 oblong leaflets. Flowers white or pink, borne solitary or in short axillary racemes; calyx 5-toothed; pod many-seeded, longer than calyx.

BIRDSFOOT FENUGREEK   *Trigonella purpurascens*

A slender, prostrate annual herb, distinguished from Clover by the pods which are longer than the calyx. Leaves 3-foliolate like a bird's foot. Flowers pinkish-white, borne 1–3 from axils of leaves on stalks shorter than those of Clover. The flowers have the soft sweet scent of White Clover while the leaves, when drying, emit the refreshing scent of newly mown hay, due to the presence of coumarin as in the closely related *Melilotus.* In bloom July–August.

Habitat: Present in dry, sandy places throughout the British Isles, especially in S England and generally close to the sea.

History: The plant takes its name *Trigonella* from the 3-angled corolla and its country name from *foenum-Graecum,* Greek hay, for the plant was used in S Europe, where it abounds, to put with hay of inferior quality. It probably accompanied the Romans with their invasion of Britain for the same purpose.

MELILOTUS (Melilot) Tall-growing annual or biennial herbs with trefoil leaves divided into narrow leaflets. Stipules joined to

Plate 7
Pea Family   1 Tree Lupin   2 Gorse

Plate 8
Rose Family (i)   1 Meadowsweet   2 Common Agrimony
                      3 Fragrant Agrimony 4 Salad Burnet

petiole. Flowers yellow or white, borne in drooping racemes. Calyx 5-toothed. Pod blunt and straight, brown or black when ripe. The foliage releases the smell of newly mown hay when drying, which is due to the presence of coumarin or coumaric acid ($C^9H^6-O^2$). Coumarin is white with a boiling point of 290° C and crystallises into rectangular prisms. It is present in, and has the same smell as, Tonquin Beans, the seeds of *Dipterix odorata*, native of the tropical forests of Guyana. The black seed is finely ground and used in dry perfumery for sachet powders and in snuffs. Coumarin is present also in Sweet Vernal Grass and in Woodruff. In bloom July–August.

COMMON MELILOT *Melilotus officinalis*

An erect branched biennial growing 3 ft tall with pale green trifoliolate leaves. The flowers are borne in lax one-sided racemes and are yellow, with the keel shorter than the wings and standard. The stigma projects beyond the anthers making self-fertilisation impossible but the calyx is short and wide, thus allowing a wider variety of insects to reach the honey. Scent is therefore unnecessary for attracting the longer-tongued insects, which is more usual in the Legumes. In bloom July–August.
Habitat: Common on waste ground, especially on grassy banks in S England and S Ireland.

WHITE MELILOT *M. alba*

An erect biennial growing 4 ft tall. Its leaves have longer leaflets than *M. officinalis* while its flowers are smaller and are white, with the standard longer than the wings and keel. It has coumarin-scented leaves. In bloom July–August.
Habitat: Common on waste ground in S and W England and Wales; less common elsewhere.

TRIFOLIUM (Clover) Herbs with ternate leaves, the stipules adnate to the petioles. Flowers sessile or sub-sessile, borne in heads or spikes, red, pink or white (rarely yellow). The clover is an example of the degree of colour (pigment) in a flower replacing scent. The flowers of the White Clover, *T. repens*, have a rich sweet perfume and its reproductive organs are so arranged as to exclude those insects with

White Melilot

short tongues. Darwin, however, discovered its need for insect fertilisation, for when he covered the flowers with muslin, he found that the plant was only one-tenth as productive as when long-tongued insects were admitted. With the Red Clover, e.g. *T. pratense*, which by its floral mechanism admits a larger variety of insects, the scent is less pronounced, while in the crimson-flowered *T. incarnatum*, scent is almost non-existent. After fertilisation, the White Clover almost immediately loses its scent, hanging its florets and turning brown. The scent from a field of White Clover when the flowers are freshly open in June is almost as powerful as that of a field of Beans when in bloom.

The calyx is 5-toothed; the petals persistent, with the wings longer than the keel. The pod is small and almost enclosed by the calyx. Honey begins to flow about 10 days after the first flowers open. Those appearing towards the end of the summer produce little nectar. Each flower-head may contain as many as 100 florets. In bloom May–August.

RED CLOVER *Trifolium pratense*
So called, even though its flowers are pink, sometimes with a purple flush. A downy perennial, the leaflets are broad and sometimes notched, with a white crescent or 'horse-shoe' on the upper surface, and stipules terminating in a long bristle. The flowers are borne in terminal, almost oblong heads. The calyx is hairy with teeth about half as long as the corolla. It is the earliest flowering Clover. In bloom mid-May onwards.
Habitat: In grassland. It is an important part of the hay crop throughout the British Isles.
History: The secretion of honey is even greater than in the White Clover. Country children call them *honeysuckles*, but the honey is not so

Red Clover

freely produced at such low temperatures as in the White Clover, nor does it rise to a level at which it is as readily available to the bee. An extract obtained from the flowers was at one time thought to have powers of curing cancerous sores, possibly due to the high lime content of the plant. Countryfolk used to make a syrup for children to ease their whooping cough.

WHITE OR KENTISH CLOVER *T. repens*

With *T. fragiferum*, it is our only prostrate species, rooting from the leaf-nodes. It can be recognised by the small white mark at the base of the leaflets. The long-stalked flowers are white, sometimes tinted with pink, and appear in globular heads with an abundance of nectar secreted at the base of the stamens. The calyx-tube is white, veined with green and with triangular teeth. The 4-seeded pod is almost hidden in the calyx. In bloom June–August.

Habitat: In pastureland and other grassy places, particularly in limestone regions, as in Kent, the Cotswolds and Chilterns.

History: It may have been the leaf of a form of *T. repens* which St Patrick chose to illustrate the doctrine of the Trinity. It takes its country name from the Latin *clava*, a club, and the 'clubs' of playing cards represent Clover leaves. From the contentment of cattle in a field of Clover has come the phrase *living in clover*. The native White or Kentish Clover differs from the so-called Dutch strain in that it is of creeping habit and is longer-living. The seed is more expensive and is mostly obtained from the old sheep-grazing pastures of Kent. It is one of the most valuable of all plants used by bee-keepers, who are of the opinion that the native form is more readily worked than the Dutch variety. In addition to the quantity of honey secreted, the wild White Clover secretes its nectar at lower temperatures than any other native plant, before the temperature rises to a point where bees will take flight. The US Department of Agriculture's Bulletin No. 1215 states that in its view, the flow of honey is most prolific when the night temperature falls just below 65° F and the day temperature is just above that point.

Unlike almost all other honeys which have a musky flavour and aroma, clover honey is of light amber colouring, with a delicate, pleasing aroma.

ASTRAGALUS (Milk-vetch) Perennial herbs or shrubs often with thorns, present on steppes and prairies of the world (except Australia). Leaves imparipinnate, stipulate. Flowers borne in axillary racemes or spikes. Calyx with 5 sub-equal teeth; corolla with blunt keel. Pod 2-valved; seeds 2-numerous. Flowers not scented, bee pollinated. In bloom June–August.

WILD LIQUORICE *Astragalus glycyphyllos*

A prostrate, perennial glabrous shrub sending out its stems to 3 ft or more. The leaflets are in 5–6 pairs and are sweetly scented. The flowers are creamy-white, borne in short racemes and contain nectar. They are followed by a smooth, curved pod. In bloom June–August.

Habitat: Present throughout England, Scotland and Wales but not in

Ireland. Common in N England, usually inhabiting limestone mountainous
slopes and waste ground.

VICIA (Bean) Climbing or non-
climbing annual or perennial herbs of
which the Persian or Field Bean, *V.
faba*, has been grown in Britain as a
fodder crop since Roman times and is
naturalised around farm fields. The
Broad Bean and the Field Bean are
varieties of *V. faba*. Many, including
the Vetches, are herbs climbing by tendrils. The flowers are borne in axillary
racemes, the wings joined to the keel.
The stamens diadelphous; the cylindrical style, thread-like with a tuft of hairs
near the end. Only those bees and
Lepidoptera with long proboscis can
reach the deeply secreted honey of *V.
faba*; hence it is the only species with
scented flowers.

Wild Liquorice

FIELD OR HORSE BEAN *Vicia faba*

An erect annual without tendrils and with square stems. Acclimatised
in Europe and the British Isles where it has been grown from earliest
times. It attains a height of 2–3 ft and bears its flowers in axillary
clusters. The flowers are white with a velvet-black blotch and appear in
June, though time of flowering depends upon whether the seed is sown
in autumn or in spring. The flowers are followed by green pods 6 in.
long containing large seeds. After Clover, it is one of the most valuable
sources of honey, which is of dark amber colouring with a pleasant
scent and taste.

Habitat: Often present on the borders of cultivated land and in hedgerows as escapes from the farm crops.

History: A single flower inhaled in the hand has a perfume which is
almost imperceptible and faintly sweet, but a field of the plants in full
bloom emits an almost intoxicating scent, detectable from more than a
mile away on a warm day with a gentle breeze.

> Sweet is the blossom'd bean's perfume,
> By morning breezes shed.

The heavy perfume of the Bean has been known to cause fantasies
during sleep and this is particularly so when one rests close to a field of

plants in full bloom. The heavy, balsamic scent, like the perfume of the Oriental Hyacinth, is similar to that given off by the Tobacco Plant and several species of tropical Orchid and other plants of tropical climes. Shenstone has written that:

> The odour of the bean's perfume,
> Be theirs alone who cultivate the soil . . .

It is indeed one of the most powerful of all flower scents and, like all wild flower perfumes, is to be enjoyed by those who work in the woodlands and about the countryside.

'Sweet I scent the blossom'd bean', wrote John Clare, in his lines 'To Summer Morning', and later in the same poem he describes the plant's pollinating insects:

> . . . the bee at early hours
> Sips the tawny bean's perfumes;
> White butterflies infest the flowers,
> Just to show their glossy plumes.

From Chaucer's time, the Bean has been used as a symbol of contempt —as we find in *The Merchant's Tale*: 'None other lif, say'd he, is worth a bene'.

ROSE FAMILY Rosaceae

Trees, shrubs, or herbs mostly scattered over northern temperate regions of the world, with alternate leaves, usually stipulate; the stipules often adnate to the petioles. Thorns present in genus *Rosa*. Flowers hermaphrodite, borne in racemose or cymose inflorescence. Sepals 5, united; petals 5, white or dingy-coloured, alternate with sepals. Stamens 2–4 times number of petals. Carpels 1-numerous, free or united to each other; styles free, occasionally united. Fruit, an achene, drupe or pome.

Associated with the genus *Rosa* which contains some of the most exquisitely scented of all flowers, it is surprising to find in this family some of the most evil-smelling of flowers. They are generally either devoid of nectar or make it readily available to the shorter-tongued insects in shallow cups and pollination is mostly by flies. The flowers are of the Aminoid Group, containing trimethylamine and propylamine which produces the unpleasant ammonia and fishy smell in the flowers of *Cotoneaster, Amelanchier, Sorbus, Pyrus communis, Crataegus monogyna* (Hawthorn), though in the latter the fishy smell is counteracted by the presence of anisic aldehyde. This is chemically related to anethol, which is the substance that gives the familiar sweet smell to aniseed. It is also present, without trimethylamine, in the flowers and roots of the Cowslip, *Primula veris*.

The clustered flowers also tend to turn towards the light sideways and in default of insect visitation, self-fertilisation occurs when the pollen falls on the stigma, as in the Blackthorn, *Prunus communis*, *Geum, Potentilla*, etc. Neither smell nor scent is noticeable in the flowers and is not necessary, as the flowers are provided with their own powers of reproduction. It would seem that no member of the Rosaceae family is visited by Lepidoptera, while visits from Hymenoptera are rare.

FILIPENDULA   Glabrous perennial herbs growing 6 in. to 4 ft tall with radical stipulate leaves. The stems arise from a tuberous rootstock. Flowers are borne in a cymose inflorescence; petals and sepals 5–8; stamens 20–40; carpels 5 or more. Fruit, a ring of follicles. The flowers, which contain no nectar, are visited by short-tongued insects attracted by the crowded inflorescence. In the absence of insects, self-fertilisation occurs. In bloom May–September.

MEADOWSWEET, QUEEN OF THE MEADOWS, BRIDEWORT   *Filipendula ulmaria*   Pl. 8

Syn: *Spiraea ulmaria*

A glabrous perennial growing 3–4 ft tall with wiry furrowed stems and pinnate leaves, shining dark green above, grey on the underside, the hairs releasing a refreshing fruity scent when handled. The terminal leaflet is lobed while the leaflets have serrated edges. The flowers are creamy-white and are borne in crowded cymes. They secrete no honey and are visited only by pollen-seeking insects, self-fertilisation taking place in their absence. Trimethylamine is present, though in less concentration than in *Crataegus, Sorbus*. The 6–10 carpels, erect at first, later become twisted in a coil. In bloom June–September.

Habitat: By rivers and ponds and about low-lying ground such as marshland and meadows; also in damp woodlands (not in acid soils) throughout the British Isles.

History: The Dutch named it *Reinette, Queen of the Meadows* or *Little Queen,* for in Holland and in Britain until recent times no plant was held in greater esteem. Ben Jonson, friend of Shakespeare, called it *Meadow's Queen*:

> Bring too, some branches forth as Daphne's hair,
> And gladdest myrtle for the posts to wear,
> With spikenard weaved and marjorams between
> And starr'd with yellow-golds and meadow's queen.

John Clare in his poem 'To Summer' wrote:

> The meadowsweet taunting high its showy wreath,
> And sweet the quaking grasses hide beneath.

It was Queen Elizabeth I's favourite strewing plant, but for its aromatic foliage rather than for the sickly hawthorn-like scent of its flowers.

Those who smell the flowers rarely press the leaves to release the invigorating sharp aromatic scent, which is due to the presence of oil of wintergreen as in *Gaultheria procumbens*. Gerard tells us that the leaves 'far excel all other strewing herbs, to deck up houses, to strew in chambers, halls and banqueting houses in summertime; for the smell thereof makes the heart merry and delighteth the senses: neither doth it cause headache as some other sweet smelling herbs do'. Parkinson described it as having a 'pretty, sharp scent' and adds that 'a leaf or two layd in a cup of wine, will give as quick and fine a relish there to as Burnet will'. Michael Drayton in the *Polyolbion* wrote:

> Amongst these strewing herbs, some others wild that grow,
> As burnet, all abroad, and meadow-wort they throw.

Like Fennel, the distilled water of the leaves was used to strengthen the eyes and to prevent itching. Both the leaves and flowers make a welcome addition to a pot-pourri, retaining their strength for many months, while the fragrance of the flowers surprisingly becomes more and more pleasant with age. The leaves also impart a delicious flavour to soups. It is a quite delightful plant to grow by the side of a stream or pond for use in pot-pourris, and is propagated by division; the plants can also be raised from seed. It takes its name *ulmaria* from the Latin *ulmus*, elm, from the shape of its leaflets.

FRAGARIA (Strawberry)
Perennial herbs increasing by stoloniferous runners. Leaves ternate, stipules adnate to petiole. Sepals 5; bracteoles 5. Flowers borne in axillary cymes. The honey secreted by a narrow fleshy ring at base of the tube, between the stamens and outer carpels. The flower opens flat, to provide a suitable landing place for the short-tongued insects. No scent is present in the white flowers. Fruit, achenes sunk in a fleshy receptacle; fragrant.

WILD STRAWBERRY
Fragaria vesca
An almost prostrate perennial, increasing by runners or by division. The leaves are trefoil, bright green and hairy and release a pleasant musky smell

Wild Strawberry

when dying in autumn. The natural oil extracted from the leaves is similar to that of *Betula lenta*, known as *Russian leather*. The white flowers are borne on a hairy peduncle and have 5 petals, overlapping. They are followed by small yellow fruits flushed with red and held on drooping pedicles with the calyx reflexed. In bloom April–July.

Habitat: With his detailed knowledge of the plants of the countryside, Shakespeare in *Henry V* described where to find the Wild Strawberry:

> The strawberry grows underneath the nettle,
> And wholesome berries thrive and ripen best
> Neighbour'd by fruit of baser quality.

It is widespread throughout the British Isles in deciduous woodlands and beneath hedgerows. It likes shade and moisture at its roots.

History: The Earl of Gloucester in Shakespeare's *Richard III* referred to the Wild Strawberry, this being the first reference to its garden cultivation. Says Gloucester:

> My Lord of Ely, when I was last in Holborn,
> I saw good strawberries in your garden there.

The gardens of the Bishops of Ely were situated where Ely Place and Hatton Garden, Saffron Hill and Vine Street now lie. They covered 40 acres on the banks of the upper Fleet, only a short distance from Gerard's garden off Fleet Street. The Ely gardens were, in Elizabethan times, given to the Queen's favourite, Sir Christopher Hatton, who had much coveted them. In return, Bishop Cox of Ely was to be permitted to walk in the gardens and gather 20 bushels of Roses a year.

The plant takes its familiar name from the straw which was placed between the rows to protect the fruit from soil-splashing, and its botanical name from the Latin *fraga*, a berry. It may have given the word *fragrance* to the English language, either from the smell released by the dying leaves or from the fragrance of its fruit. Thomas Tusser said that:

> If frost do continue, take this for a law,
> The strawberries look to be covered with straw.

By early Stuart times, strawberry-growing (and this was the wild species, for the South American *F. chiloensis* had not yet reached England) had produced a profitable crop around London, the fruit being transported in large wicker baskets, carried on the heads of the women. Southgate was the most popular growing area and the early fruit sometimes made as much as £10 a pound, an enormous price in those days.

GEUM Pubescent perennial herbs growing from a thick rhizomatous, sometimes aromatic, rootstock. The leaves pinnate with 2–3 pairs of lateral crenate leaflets. The flowers borne in small scapes; sepals and petals 5; carpels numerous. Fruit, a group of achenes.

HERB BENNET, WOOD AVENS *Geum urbanum*

Herb Bennet

An erect, hairy perennial growing 12–20 in. tall; its radical leaves pinnate, its cauline leaves ternate. The flowers have 5 petals as long as the sepals and measure about ¾ in. across, being smaller than those of *G. rivale*. They are yellow with no scent. Honey is secreted by a fleshy ridge situated at the base of the tube. When the flower opens, the stamens bend inwards so that the anthers lie near the outer carpels while the inner red styles with their ripe stigmas project towards the centre. Self-fertilisation thus occurs in the absence of insects (mostly flies and beetles) and their visits are usually few. In bloom June–August.

Habitat: Widespread in deciduous woodlands and on grassy, shaded banks throughout the British Isles but now becoming less common. Prefers shade and moisture at its roots.

History: Its country name is *Herb Bennet* or the *Blessed Herb*. It is also the *Holy Herb*, as the 5 petals represented the 5 wounds of Christ and its medicinal properties were considered valuable, while the roots were thought to be a charm against superstitions.

Mary Quelch in *Herbs and How to Know Them* (Faber) has said that 'the roots, tied in small bundles and placed in an apple tart will take the place of cloves, imparting a delicious warm, aromatic flavour'. Indeed, it takes its botanical name from the Greek *geuo*, I taste. Culpeper also mentioned the clove-like fragrance, especially when the roots were dry. During his time and at a much earlier date, they were placed in ale and wine (as a substitute for Pinks) to impart their clove-like flavour. The foliage, too, is clove-scented and as Culpeper said, 'is used to expel crude and raw humours from the belly by its warm savour'. A hot cordial made from the leaves provides a warming drink on a cold day.

WATER AVENS *G. rivale*

A downy perennial herb growing 18 in. tall and differing from *G. urbanum* (p. 89) in its smaller stipules, greater degree of pubescence and its drooping flowers. The radical leaves have 3–6 pairs of leaflets; the cauline leaves being few, with few leaflets. The flowers are about twice the size of those of *G. urbanum*, are orange-yellow with pink veining and with large reddish-brown sepals; the styles persistent, hooked. In bloom May–June.

Habitat: Common on marshy ground and in ditches and by streams throughout Scotland and N England; rare elsewhere; absent from SE England.

History: In the USA it is known as *Chocolate root* for its roots when dried are purple-brown and were at one time used as a substitute for cocoa. In Britain, the clove-scented roots were used to make a tonic beer. To quote Culpeper, 'the root in springtime steeped in wine, doth give it a delicate taste . . . whilst it is a good preservative against the plague'.

AGRIMONIA Perennial herbs with stipulate, pinnate, serrated leaves, slightly glandular. Flowers yellow, borne in spike-like racemes. Sepals 5, persistent; petals 5; stamens 12–20; carpels 2. Fruit composed of 1–2 achenes. In bloom June–July.

COMMON AGRIMONY *Agrimonia eupatoria* Pl. 8

A slender perennial growing 2 ft tall with reddish stems. The lower leaves are composed of 3–6 pairs of leaflets with toothed edges. The small yellow flowers are borne on spike-like racemes, the calyx-tube being deeply furrowed, with hooked spines at the mouth. The flowers emit a faint but definite scent of ripe apricots, as do the roots. Culpeper said 'it hath a reasonable scent' and he recommended a concoction of the leaves and roots for kidney troubles. Pollination is by flies and bees. In bloom July–August.

Habitat: Common on the borders of fields and in hedgerows and on waste ground throughout the British Isles.

FRAGRANT AGRIMONY *A. odorata* Pl. 8

Closely related to *A. eupatoria* (above), but of more branching habit and growing taller, while the leaf scent is more pronounced, being resinous though lightened with a sweetness, similar to the leaves of the Walnut tree. Borneol acetate is present in the leaves (as in conifer leaves) and imparts the familiar turpentine smell. The stems are more leafy and the leaves larger, with the teeth more deeply defined. The flowers too, are larger and of a deeper yellow colour and have much the same aromatic quality as the leaves. The flowers are borne closer together along the stem and have a bell-like calyx-tube. In bloom June–July.

Habitat: Common in SW England and in S Wales but now rare elsewhere in the British Isles. Usually found in hedgerows and on grassy banks but not in calcareous soils.

History: Countrymen used to make a tonic beer from the leaves, while the tendency of both flowers and leaves to retain their aromatic scent long after drying made them very popular for making pot-pourris and for stuffing cushions, pillows and mattresses.

POTERIUM Erect perennial herbs with pinnate leaves, the stipules adnate to the petiole. Flowers green or purple-brown, borne in a capitate cyme on a long peduncle. Calyx-tube persistent, sepals 4; petals absent. Stamens 4–30 with long slender filaments which protrude from the flower. Flowers without nectar and wind-pollinated; scentless, though the leaves are fragrant. Achenes solitary, enclosed in calyx-tube.

SALAD BURNET *Poterium sanguisorba* Pl. 8

An almost glabrous erect perennial growing about 12 in. tall, the leaves comprising 4–12 pairs of oval leaflets with serrated edges. They have a mild cucumber-like smell and taste, resembling Borage. The tiny flowers are green, tinted purple and are borne in a capitate cyme, the uppermost flowering being female; these form their red stigmas before the lower flowers form their yellow stamens. The calyx tube is 4-winged. In bloom May–August.

Habitat: Common in England, Wales and Ireland, localised in Scotland. Present in fields and grassy banks, usually in a calcareous soil. Especially common in S England, NE Yorkshire, W Ireland.

History: Both Parkinson and Culpeper said that the leaves were used in salads (they are delicious with cheese, in a sandwich) and to flavour claret and other drinks, imparting a pleasant 'quick' taste. From its use in this way, Pliny tells us that it obtained its name from the Greek *poterion*, a drinking cup. Michael Drayton described his liking for the leaves:

> The burnet shall bear up with this
> Whose leaf I greatly fancy.

PRICKLY SALAD BURNET *P. polygamum*

Syn: *P. muricatum*

A taller, leafier form than *P. sanguisorba*, it is perennial and may have been introduced by the Romans, for it abounds in Italy and S Europe where it is in great demand as a fodder crop. Turner in his *New Herbal* admirably described it as having 'two little leaves (deeply toothed) like unto the wings of birds . . . setteth out when they intendeth to fly'. The leaflets are borne 5–12 to a stem, directly opposite each other, like the wings of a bird in flight. The flower-heads are oval with prickly wings to the calyx-tube.

Habitat: Though possibly an introduction, it is completely naturalised around fields and on downlands especially in S England, the Midlands (Cotswolds), and in Wales. Rare elsewhere.

History: This was the plant which Purten, in his *Flora of the Midland Counties,* said formed almost the whole of the vegetation on the sheep walks of Wiltshire in sight of Salisbury Cathedral spire. It abounds in Europe and is the plant which Shakespeare, who must have known it well from the sheep walks of the Cotswolds, coupled with those other sweetly scented plants, the Cowslip and Clover, in the Duke of Burgundy's famous speech made to Henry V and the French king:

> The even mead that erst brought sweetly forth
> The freckled cowslip, burnet and green (white) clover
> Was now growing instead, the deadly hemlock.

This was the Burnet which was widely planted in garden 'walks' during Tudor times, with Chamomile and Thyme. Francis Bacon said that the plants which 'do perfume the air most delightfully, not passed by as the rest but being trodden upon and crushed, are three – Burnet, Wild Thyme and Water mint'. Culpeper said that it was to be found 'near London, by St Pancras Church and . . . in a field by Paddington'.

ROSA Deciduous shrubs with epidermal appendages (thorns); the leaves pinnate, serrate, stipulate. Stipules adnate to sheathing petiole. Flowers borne terminal, solitary or in cymes. Calyx-tube persistent with 5 leafy, imbricate sepals. Petals 5; stamens numerous, inserted on disc at mouth of calyx-tube; carpels at bottom of tube. Fruit (hips), a collection of achenes. The flowers mostly possess a refreshing fruity perfume and are pollinated by beetles, flies and occasionally Hymenoptera. The flowers secrete no nectar but have an abundance of pollen supplied by the numerous stamens. Rose scents are so varied as to cover every group of flower scents. The true old rose scent is present in *R. gallica* and *R. damascena*; the lemon scent in *R. bracteata*; the musk scent in the Field Rose, *R. arvensis*; the scent of mignonette in *R. canina*. The heavy, unpleasant smell due to the presence of indol, one of the substances of putrefaction, is found in the Austrian Briar, *R. foetida*, its flowers smelling like those of *Lilium candidum* inhaled near to but with the fishy undertones due to trimethylamine, present in the Cotoneaster and Hawthorn of the same family.

FIELD OR MUSK ROSE *Rosa arvensis* Pl. 9

A glabrous species with purple stems which trail downwards and are covered in hooked thorns. The leaflets are in 2–3 pairs with serrated edges and with hairs on the veins. The flowers are borne solitary or in groups of 2–6 and are white with reflexed purple sepals. Styles united in a column. Fruit, a small hairless hip. In bloom July–August.

Habitat: Common throughout the British Isles (rare in Ireland) in deciduous woodlands and hedgerows, especially in S England; now more rare in the north.

History: It is Shakespeare's 'musk rose' of *A Midsummer Night's Dream*, while Bacon included it with those flowers 'yielding the sweetest smell in the air', though possibly both had in mind the Musk Rose, *R. moschata*, of the Himalayas, introduced into Britain during the reign of Henry VIII. At the turn of the century, however, it would still have been rare and it is more likely that the reference was to the native Field Rose, a familiar plant of the Warwickshire hedgerows which emits a musk or honey-like scent from its flowers. This is especially noticeable on damp, warm days when it perfumes the air for some distance. This may have been the flower in Shakespeare's thoughts when in *Henry VI, Part 2*, Richard Plantagenet, Duke of York, makes his famous speech condemning the action of the Duke of Suffolk who has returned with the King's bride-to-be, Margaret of Anjou. By the marriage settlement, Anjou and Maine go to her father, though Margaret brings no dowry in return:

> Till Henry, surfeiting in joys of love,
> With his new bride and England's dear-bought queen,
> And Humphrey with the peers be fall'n at jars:
> Then will I raise aloft the milk white rose,
> With whose sweet smell the air shall be perfum'd,
> And in my standard bear the arms of York,
> To grapple with the House of Lancaster;
> And force perforce, I'll make him yield the crown
> Whose bookish rule hath pull'd fair England down.

The Field Rose fades so quickly when plucked that it was possibly the Rose mentioned by Othello when he enters the bedchamber of Desdemona planning to kill her, although he refrains from doing so:

> When I have pluck'd thy rose,
> I cannot give it vital growth again,
> It needs must wither: I'll smell thee on the tree.

The Field Rose may also have been the White Rose of the Temple gardens where Plantagenet and Somerset begin their fatal quarrel:

Plantagenet: Let him that is a true-born gentleman,
               And stands upon the honour of his birth . . .
               From off this Brier pluck a white rose with me.

Somerset: Let him that is no coward nor no flatterer
               But dare maintain the party of the truth,
               Pluck a Red rose from off this thorn with me.

It is possible that the double White Rose (of York) to be found in gardens of the time was, in fact, the double form of *R. arvensis*, for Gerard said that 'the double White Rose doth grow wild in many hedges' and he singles out Lancashire and the northern counties as the places in which it was most common. Pliny the Elder, discussing the derivation of the word Albion (the ancient name for Britain), believed it to come from the White Rose, *Rosa alba*, 'with which it abounds'. It may be that the several lovely varieties of the White or Alba Roses are descended from the double form of the Field Rose, for most are of ancient origin and are highly scented. Gerard in his *Herbal* said that 'the rose deserveth the chiefest and most principal place amongst all flowers whatsoever, being not only esteemed for its beauty, virtuous and fragrant and odoriferous smell but because it is the honour and ornament of our English sceptre'.

In his *Delights for Ladies* (1594), Sir Hugh Platt, 'knight of Lincoln's Inn, gentleman', gave details for retaining the scent of rose-water, and for making pot-pourris he recommended that you should 'hang your pot in an open chimney or near a continual fire so that the petals will keep exceeding fair in colour and most delicate in scent'.

In the *Ashmolean Manuscripts* there is a recipe for making perfume for Henry VIII to be used about his person and on his clothes. 'Take six spoonfuls of rose oil and six of rose water and a quarter of an ounce of fine sugar. Mix and add two grains of musk and one ounce of ambergris. Boil together softly for 5–6 hours and strain'. In *A Queen's Delight* (1662) is a recipe for making 'An Odiferous Perfume for Chambers' which consists of heating together powdered cloves and rose-water. Mary Doggett in her *Book of Receipts* (1682) suggests warming Rose leaves with orris powder to scent a room and also gives details of how the scent of Damask Roses may be retained, to be used for placing with linen. Indeed, Roses were grown in large numbers, for the flowers possessed the same properties as herbs. Besides the pleasing fragrance of the bloom, syrup of roses was suggested as a pick-me-up for the winter months as long ago as 1550, when a recipe appeared in Ascham's *Herbal*. 'A conserve of roses . . . taken in the morning fasting and last thing at night, strengthens the heart and taketh away the shaking and trembling', wrote Gerard and he also advocated the use of distilled rose-water for the same purpose. 'It . . . (also) bringeth sleep, which also the fresh roses themselves provoke through their sweet and pleasant smell'.

On the petals of the more highly scented Roses, minute perfume glands on the upper surface may be clearly seen through a microscope. These contain the volatile oil. With the Moss Rose, a 'sport' from the old Cabbage Rose, R. *centifolia*, the scent glands on the flower stems and on the sepals (which enclose the buds) have become so enlarged as to be clearly visible to the eye in the form of 'moss' which has the fragrance of the bloom. The same glandular hairs are present on the leaves of the Sweetbriar which, when gathered, will leave their fragrance on the hands for a considerable time. Father Catron, in his *Historie de Mogol*, tells how the Princess Nourmahal caused a large tank, in which she used to be rowed with the Great Mogul, to be filled with rose-water. The sun's heat caused the essential oil to be separated from the water, it floated on the surface and was collected and used as perfume. Indeed, in parts of India and the Middle East, this method of collecting the essential oil is still practised. Rose-water is placed in large open jars to enable the sun to separate the essential oil; this is then removed by skimming and placed in bottles. Near the homes of the nobility, large plantings of rose-bushes are made rather like our orchards, so that the fragrance of the blooms permeates the whole house, while the plants provide pleasure during the evening and early morning when, fresh with dew, they emit a powerful and most refreshing fragrance when one walks among them.

BURNET OR SCOTTISH ROSE R. *pimpinellifolia*
Syn: *R. spinosissima* Pl. 9

A low-growing multi-branched shrub, increasing and spreading to form large thickets by means of underground suckers. The stems grow to a height of 2 ft or more and are covered in long spines, passing into glandular hairs. The leaves are pinnate, with 7–9 serrated spines and sometimes slightly hairy on the underside. The flowers are borne solitary and are without bracts. They are creamy-white with a delicate fruity scent and are followed by crimson-black hips. The shrub was known to Tudor gardeners. In bloom May–July.

Habitat: On sand-dunes and coastal areas in Scotland (hence its name), especially in the west; also in Ireland. Rare in England and Wales. Two varieties of garden value are Falkland, found in the grounds of the Royal Palace of Falkland in Scotland and which has grey-green foliage and bears semi-double blooms of pale pink; and Stanwell Perpetual, possibly the result of a chance crossing with the Autumn Damask, for it is perpetual flowering. It was discovered in a garden at Stanwell in Middlesex and was introduced into commerce by

Lee of Hammersmith. Its bluish-white flowers, borne on graceful arching stems, have a rich sweet perfume.

DOG ROSE, WILD ROSE, ENGLISH ROSE
R. canina   Pl. 9

The commonest wild Rose of the British Isles, it forms graceful arching stems covered in large hooked prickles shaped like a dog's tooth, and wide at the base. The leaves are composed of 2–3 pairs of serrate leaflets and are hairy about the veins. The flowers may measure 2 in. across and are pale pink, occasionally white. They are borne mostly solitary on long footstalks and are sweetly scented. The sepals are reflexed while the numerous golden stamens enhance the beauty of the bloom. At the end of the calyx-tube is a fleshy ring surrounding the styles which allows only the stigmas to protrude. In bloom June–July.

Habitat: Common in woodlands and hedgerows throughout England and Wales but less common in Scotland and in Ireland. A double pink form, Abbotswood, found in the garden of the late Mr Harry Ferguson, of tractor fame, is a delightful garden plant.

History: The Dog Rose is one of the longest living of all plants, and the one growing in a convent garden at Hildesheim in Germany is said to have been planted there by a son of Charlemagne in AD 850. Its massive roots are almost indestructible and a distilled water was obtained from them which was believed to be a cure for rabies. Shakespeare knew it as the *Canker Rose* and always spoke of it in derogatory terms. In *Henry IV, Part 1*, Bolingbroke has usurped Richard's throne and made himself king. The Canker was a wildling, the Rose of the hedgerow never of the garden, and in the play Hotspur speaks in anger at the thought of Henry seizing the throne in place of Mortimer, the rightful heir:

> To put down Richard, that sweet lovely rose,
> And plant this thorn, this canker, Bolingbroke?

In *The Sonnets* Shakespeare again compares the Dog Rose with the Red Gallica Rose, the Apothecary's Rose, whose rich perfume is retained long after the petals are dried and made into pot-pourris, while the flower of the Dog Rose loses its scent immediately after fertilisation:

> The Canker blooms have full as deep a dye
> As the perfumed tincture of the roses,
> Hang on such thorns and play as wantonly
> When summer's breath their masked bud discloses;
> But, for their virtue only is their show,
> They live unwoo'd and unrespected fade;
> Die to themselves – sweet roses do not so;
> Of their sweet deaths are sweetest odours made.

Plate 9
Rose Family (ii) 1 Sweetbriar 2 Field Rose
3 Burnet Rose 4 Dog Rose

Plate 10
Rose Family (iii) 1 Hawthorn   2 Rowan
3 Pear

The Dog Rose is the 'conventional' heraldic Rose and was usually shown 'seeded on' with the green leaves (calyx) appearing between the petals. In the grant of arms to William Cope, cofferer to Henry VII, the Roses are shown 'slipped', with a stalk added. A single Dog Rose constitutes the arms of Montrose.

SWEETBRIAR, EGLANTINE   *R. rubiginosa*   Pl. 9

An erect dense bush with its tall, arching stems clothed in hooked spines. The leaves are doubly serrate, downy and with brown glands on the underside which release an aromatic fruity scent like ripe apples when pressed, or even without bruising when the sun shines upon them. This is most pronounced when sunshine and rain alternate. The scented flowers, borne solitary or 2–4 together, are usually deep rose-pink, occasionally blush-white. They are followed by brilliant scarlet pear-shaped hips which are retained well into winter. In bloom May–July. *Habitat:* Widespread in the Midlands (the Cotswolds) and S England, less common in the north; rare in Scotland and in Ireland. Mostly in hedgerows and where the soil is of a calcareous nature.

Two distinctive forms make attractive garden plants – Janet's Pride, discovered in a Cheshire hedgerow in 1892, the deep pink flowers having a striking white centre; and La Belle Distinguée, the Double Scarlet Sweetbriar of French origin. It makes a compact bush and bears flat semi-double blooms of numerous, crimson-scarlet petals. *History:* Oberon in *A Midsummer Night's Dream*:

> I know a bank whereon the wild thyme blows,
> Where ox-lips and the nodding violet grows;
> Quite over-canopied with luscious woodbine,
> With sweet musk-roses and with eglantine:
> There sleeps Titania . . .

Arviragus in *Cymbeline*:

> With fairest flowers,
> Whilst summer lasts, and I live here, Fidele,
> I'll sweeten thy sad grave: Thy shalt not lack
> The flower that's like thy face, pale primrose; nor
> The azur'd hare-bell, like thy veins; no, nor
> The leaf of eglantine, whom not to slander,
> Outsweeten'd not thy breath.

The earliest gardens were surrounded with Quickthorn or with Eglantine, for both were extremely durable and formed a dense, spiny, impenetrable hedge which kept out all intruders including cattle. The name Eglantine comes from the Latin *eguleius* or *aculeius*, prickly. Chaucer wrote:

The grene herber
With sycamore was set and eglatere.

'Sweet is Eglantine', wrote Edmund Spenser and Leigh Hunt also sings
its praises in the 'Chorus and Songs of the Flowers':

Wild-rose, Sweet-brier, Eglantine,
All these pretty names are mine,
And scent in every leaf is mine,
And a leaf for all is mine,
And the scent – oh, that's divine!
Happy-sweet and pungent-fine,
Pure as dew and prick'd as wine.

Robert Herrick asks us to:

Take this sprig of Eglantine,
Which though sweet unto your smell,
He who plucks the sweet shall prove
Many thorns to be in love.

SMALL-LEAF SWEETBRIAR *R. micrantha*

It makes a compact bush and is in all respects smaller than *R. rubiginosa*,
though it greatly resembles this Rose. It has hooked prickles and double
serrated leaves, glandular on the underside, which, when pressed,
release the fruity scent of the Eglantine. The flowers measure 1 in.
across, are of palest pink and borne on bristly footstalks. In bloom
July–August.
Habitat: More common than *R. rubiginosa* in soils of a non-calcareous
nature but present only in S England and Wales and in SW Ireland.
Found in woodlands and hedgerows and about wild coastal scrubland.

PRUNUS Small trees or shrubs with simple serrate leaves; petiole
glandular. Flowers white, borne singly or in clusters appearing with or
before the leaves; sepals and petals 5; stamens 20. Fruit, a 1-seeded
drupe. Nectar is secreted, but few bear scented flowers, for they are
equipped for self-fertilisation in the absence of pollinating insects.

BIRD CHERRY *Prunus padus*

A small deciduous tree or shrub with dark brown bark which releases
a pungent smell when removed. The elliptic or oval serrate leaves have
tufts of hairs at the axils of the veins. The flowers are white and are
borne in drooping racemes of 20 or more. The sepals are short and
fringed with glands while the petals are toothed. The stamens curve
inwards and as they unfold come into contact with the already mature
stigma so that self-fertilisation takes place in the absence of the usual
pollinators (flies and beetles). The flowers, however, have a sweet

almond-like perfume which is altogether absent in the double garden form, *rexii*, one of the loveliest of the May-flowering Cherries.

Habitat: Common in hedgerows and deciduous woodlands and on rocky ground throughout the British Isles, but especially on calcareous soils as along the south coast; in N Midlands, N Yorkshire and the Scottish border counties.

History: Present in the most exposed situations, especially in calcareous soils. Gerard said it grew wild in the woodlands of Kent and in Westmorland, while in Lancashire it was to be found 'almost in every hedge'. It is a valuable wind-break tree.

COTONEASTER Small trees or shrubs, evergreen, semi-evergreen or deciduous with entire leaves, ovate or oval. Flowers small, white, cream or pink, borne solitary or in corymbs. Pollination is by wasps or dung-flies and the flowers are among the most evil-smelling of all, containing trimethylamine, unadulterated by any other sweet-smelling substance. Fruit red or purple with 2–5 stones.

GREAT ORME BERRY *Cotoneaster integerrimus*
Syn: *C. vulgaris*

A small deciduous shrub with twigs which are tomentose when young, and with small oval leaves, downy on the underside. The minute flowers, rose-pink in colour, are borne in small cymes and have an unpleasant fishy smell. Pollination is by wasps or midges which suck the honey from the shallow cups, while in their absence self-pollination is possible. The flowers are followed by shining red globose fruits. In bloom May–June.

Habitat: Rare in rocky places across N Europe and Asia. In Britain, present only on the Great Orme near Llandudno in Wales.

CRATAEGUS Small trees or shrubs, mostly deciduous and often with thorns which are modified branches. Leaves simple, lobed; the stipules deciduous. Flowers white, creamy-white or pink, borne in terminal cymes of 12 or more. Sepals and petals 5; stamens 5–25; carpels 1–5 enclosed in calyx-tube, united at base, free at apex. Fruit, a bony pome.

HAWTHORN, QUICKTHORN, MAY
Crataegus monogyna Pl. 10

A multi-branched, small deciduous tree or shrub with smooth bark and twigs. The leaves are 3–7 lobed with occasional hairs on the veins, and taper to an acute apex. The flowers are dingy-white sometimes pink-tinged or deeper pink and possess a sweet sickly unpleasant smell like herring brine but with an undertone of aniseed. The fishy smell is due to carbon compounds of ammonia, trimethylamine and propylamine, present among the first products of putrefaction (p. 18). Pollination is by dung-flies and midges, attracted by the smell, though when the flowers are newly opening a noticeably sweeter, more pleasing scent – anisic aldehyde (aniseed) – predominates. But this is lost about two days after the opening of the flowers as they become ready for fertilisation. Immediately after pollination, the unpleasant smell disappears entirely. The anthers of the flower are purple with brown pollen which, to flies, has the appearance of decaying flesh. The flowers are followed by globose crimson fruits (haws) of 1–3 carpels. The flowers of the double red Hawthorn are devoid of any unpleasant smell. In bloom May–June.

Habitat: Common in hedgerows throughout the British Isles, for it is more widely used as a hedge than any other plant. It is especially vigorous in calcareous soils.

History: One of the oldest plants known to man, it takes its English name from the Anglo-Saxon *haive* or *haeg*, a hedge, for which purpose it was then (as today) widely used. It is also the Quickthorn, as it grows so readily in any soil. The fruits (haws) take their name from the tree. Its botanical name comes from the Greek *kratos*, strength, alluding to the hardness and durability of the wood which, like the Box, is used for engraving. In England, because of its 'deathly smell', it was always considered unlucky and is never taken indoors. It was also believed to be the tree which provided Christ's crown of thorns and Diodorus, living shortly after the death of Christ, has said that in Sicily where he (Diodorus) lived, people were buried with Hawthorn branches. Yet it has always been a much-loved tree on account of its fresh-green leaves, pleasant in sandwiches, and its mass of May blossom.

The Scottish poet James Graham wrote of 'the Hawthorn, May's fair diadem' and William Browne, contemporary of Shakespeare, described its charms in the *Britannia's Pastorals*:

> Mark the faire blooming of the Hawthorn tree,
> Who, finely clothed in a robe of white
> Fills full the wanton eye with May's delight.

By Robert Burns, it was held in greater esteem than any other tree. In his 'Farewell to Highland Mary' he wrote of the richness of the Hawthorn's blossom again, inviting one to 'breathe the milk-white thorn that scents the evening gale'.

The appearance of the Hawthorn blossom heralded the approach of summer and for this reason it brought a touch of happiness. As Robert Bridges said:

> Spring goeth all in white
> Crowned with milk-white May.

And Charles Swinburne wrote:

> The coming of the Hawthorn brings on earth
> Heaven: All the spring speaks out in one sweet word . . .

Edmund Spenser in *The Shepherd's Calendar* described the activities of the young folk who:

> . . . now flock everywhere
> To gather May-buskets and smelling Brier;
> And home they hasten the postes to dight,
> And all the kirk-pillowes eare daylight,
> With hawthorn buds, and Sweet Eglantine
> And garlands of roses and soppes-in-wine.

Upon May-morning it was the custom for every countryman to go round 'a-maying', carrying bunches of May blossom. Chaucer described how on May Day 'forth go all the Court, both most and least, to fetch the flouris freshe' and there is an account of Henry VIII a-maying with Queen Katherine of Aragon.

The tree became the badge of the House of Tudor, for it is said that after Richard's death at Bosworth Field, Lord Stanley found his crown in a Hawthorn and at once placed it on the head of his son-in-law, Henry Tudor, who left for London to be crowned the new king.

MIDLAND OR WOODLAND HAWTHORN
C. oxyacanthoides

A deciduous tree or shrub, less vigorous in growth than *C. monogyna* and with fewer thorns. The leaves are 3–5 lobed and without the hair-tufts in the veins. The flowers number 8–10 and are usually creamy-white with pink anthers. Styles 2, usually 1 in *C. monogyna*. The flowers bloom earlier and have the same smell as *C. monogyna* and are pollinated in the same way. They are followed by small crimson fruits generally containing 2 stones. In bloom early May.

Habitat: Mostly in woodlands, rarely in hedgerows, in Midlands and S and E England; uncommon in the north and in Scotland and Ireland.

SORBUS Deciduous trees without thorns; the leaves pinnate, simple, lobed or toothed. Flowers white or cream, borne in corymbs or inflorescence. Stamens 15–25; carpels 2–5, united; styles free or joined near base. Fruit crimson or yellow, usually 2-chambered. The flowers contain ammonia compounds and like those of *Crataegus* and *Cotoneaster* are unpleasant-smelling. Pollination is by flies and midges (Diptera).

ROWAN, MOUNTAIN ASH   *Sorbus aucuparia*   Pl. 10

A deciduous tree growing to a height of about 20 ft with smooth grey bark and pubescent twigs, especially when young. The leaves pinnate with 13–17 toothed leaflets. The flowers white and small and borne in flat corymbose cymes of 30 or more with cream-coloured anthers. They have an unpleasant ammonia-like smell and are followed by red or orange globose fruits (berries). In bloom June.

Habitat: Common throughout the British Isles to a height of 3000 ft or more and present in woodlands and hedgerows and about rocky outcrops in almost all, but especially calcareous, soils.

PYRUS   Deciduous trees, sometimes thorny. Leaves simple, ovate, rounded at base, serrate, tomentose when young. Stipules deciduous. Flowers in terminal cymes; stamens 20–30. Carpels 2–5, united; styles free. Fruit pyriform, dead sepals adjoining. The unpleasant smell of the flowers, similar to that of *Sorbus*, attracts Diptera as the chief pollinators. Hymenoptera may also visit the flowers but as the stamens are immature while the stigmas are ripe, insect fertilisation can take place only if the flowers are visited soon after opening. Self-pollination is possible with certain varieties of Pear but with others pollinating varieties are necessary.

PEAR   *Pyrus communis*   Pl. 10

A deciduous tree, sometimes with spines at extremities of its branches. The leaves are simple, oval, serrate, and, when young, woolly on the underside. They are held on long footstalks. The flowers are dingy-white with purple anthers (to attract flies) and are about 1 in. across. They are borne in corymbose cymes. The fruit is green or brown, 5-chambered and tapering at the base, with the dead sepals attached. In bloom May.

Habitat: Common in woodlands and hedgerows throughout England and Wales, but rare in Scotland and in Ireland.

History: 'I must have saffron to colour the Warden pies', says the Clown in *A Winter's Tale* and here the reading should possibly be *pears* not *pies*, though it may well have been a pie of Warden pears that the Clown was about to make. The Warden Pear is believed to have originated at Woburn (at one time Wardon) Abbey in Bedfordshire where it was first grown by the Cistercian monks in the twelfth century. Or it may have taken its name from the Anglo-Saxon *wearden*, meaning to keep, for this may have been a Pear with good keeping qualities, possibly introduced from France, the home of most good pears. Three Warden Pears, however, appear in the arms of Woburn Abbey; also in the arms of the County of Worcestershire where the pear was once widely grown and it was the Warden that was most commonly planted. Parkinson mentioned it, together with the Lukewards, a Pear which ripened by St Luke's Day, 18th October.

MALUS Deciduous trees, sometimes thorny and differing from Pyrus in that there are no stone cells in the flesh. The leaves are simple, toothed or lobed. The flowers are white and pink and borne in an umbellate cyme; stamens 15–50; carpels 3–5, united; styles united below. Nectar is secreted in the shallow cup and is readily accessible to all forms of insects (except for Lepidoptera) which visit the flowers. The five stigmas overtop the stamens and ripen before them, but in the absence of insects the flowers turn sideways to the sunlight so that some pollen can fall on the stigmas. A number of cultivated varieties, however, are self-sterile and it is interesting that in those which are self-fertile, e.g. Beauty of Bath, and do not need insect visitors, perfume is absent. Yet in those varieties which rely upon insect visitation for their fertility and require one or more pollinators to set a heavy crop of fruit, the scent is sweet and in certain varieties quite powerful, e.g. Bramley Seedling, Lane's Prince Albert, the most difficult to fertilise.

CRAB APPLE *Malus sylvestris*
Syn: *Pyrus malus*
A small deciduous tree, sometimes spiny with seal-brown or grey bark. The leaves are ovate or oval, rounded at the base, serrate with a 2-in. long petiole. The flowers, borne in umbels, are white, shaded pink or crimson, with golden anthers. They are followed by a globose, 5-chambered pome, yellow or green, often marked with red and with a hollow where the stalk is inserted. The calyx is persistent. In bloom May.

Crab Apple

Habitat: Common in deciduous woodlands and hedgerows throughout England and Wales; rare in Scotland and in Ireland.
History: In Holland's *Translation of Pliny*, we find 'as for wildings and crabs, . . . they carry with them a quick and a sharp smell'. It was because of this that the Crab Apple was so highly valued in medieval times for putting in ale to bring out the fullness of its flavour. Shakespeare mentions the custom of placing roasted Crabs in bowls of punch or ale. *Love's Labour's Lost* ends with Winter singing:

> When roasted crabs hiss in the bowl,
> Then nightly sings the staring owl,
> Tu-whit; to-who; a merry note,
> While greasy Joan doth keel the pot.

From the wild Crab Apple, crossed with introductions from Europe which began to arrive in Roman times, have been evolved the commercial apples of today. From earliest times, the orchard has been an integral part of monastic foundations and we know that Brithnodus, first Abbot of Ely in the seventh century, was skilled in the art of planting and grafting Apple trees. It is recorded that in King John's reign the grant of land to Llanthory Priory included 12 acres of orchards and there is a deed of 1204, relating to the holding of the lordship of the manor of Runham in Norfolk for a yearly payment of 200 Pearmains (the first named Apple) to be made to the Exchequer on St Michael's Day.

Ralph Austen writing in 1653 had this to say of the smell of an orchard: 'Chiefly the pleasure this sense (of smell) meets with is from the sweet-smelling blossoms of all the fruit trees, which from the time of their breaking forth till their fall, breathe out a most pretious and pleasant odour; perfuming the air throughout all the orchard'.

STONECROP FAMILY Crassulaceae

Perennial herbs or undershrubs, usually frequenting dry places. Many succulent, with leaves covered in wax to prevent moisture evaporation. Leaves exstipulate, simple, entire. Flowers usually small and starlike, polysymmetrical. Sepals 3–20, united at base; petals equal to sepals; also stamens or twice as many, in two whorls. Fruit a ring of follicles. Those bearing white, yellow or greenish-yellow flowers in which the nectar is readily available are visited mostly by flies, while those visited by bees, where the honey is secreted at the base of the tube, bear purple flowers. The flowers are never scented nor are they self-pollinating, for the close arrangement of the numerous flowers seems to make this unnecessary.

SEDUM Succulent perennial herbs with leaves alternate, occasionally whorled. Flowers usually borne in cymes, petals 4–6, free; stamens 8–12, in 2 whorls. Carpels free or united at base. Fruit, a ring of follicles.

ROSE-ROOT *Sedum rosea*
Syn: *Sedum rhodiola*
A grey-green succulent growing about 12 in. high with leaves broad and flat, overlapping up the erect stems. The leaves are obovate, round at the base and toothed at the apex; the lower leaves are scale-like. The greenish-yellow flowers with

Rose-root

purple anthers are borne in a terminal inflorescence, the petals being longer than the sepals, the stamens longer than the petals. Male and female flowers appear on separate plants and are pollinated by flies. They have no scent but the 3-in. long rhizomatous root, when drying, has the scent of Roses. In bloom May–August.

Habitat: Present on cliffs and rocks, especially near the sea in N Yorkshire; and from N Wales to NW Scotland; also in W Ireland and the Isle of Man.

History: No cottage garden in northern England was without its Roseroot in the past, for a distilled water was obtained from the dried roots which had a definite but diluted scent of rose-water. It was used for the complexion and was sprinkled over clothes before the advent of dry cleaning and over washing.

GOOSEBERRY FAMILY Grossulariaceae

Shrubs, evergreen or deciduous, mostly confined to North and South America and N Europe, sometimes spiny with alternate, often palmately lobed leaves, usually with stipules adnate to petiole. Flowers usually hermaphrodite or dioecious; borne solitary, more often in racemes. Sepals 4 or 5; petals shorter than sepals. Stamens equal in number to petals. Styles 2, cornate. Fruit, a berry with calyx persistent.

RIBES As above. Few bear scented flowers but *R. aureum* (also *R. odoratum*), native of North America and known as the Clove-scented Currant, has a calyx-tube 11–12 mm deep in which the honey is accessible only to the longest-tongued Hymenoptera. The flowers have an aromatic perfume, though in this case it would seem that the bees are directed more by the red veining of the petals than by the scent. Where the honey is more easily accessible, scent is absent. The leaves of several evergreen species emit a pine-like odour when pressed, due to the presence of borneol acetate, those of the native Black Currant having a cat-like smell with fruity undertones, similar to *Hypericum hircinum*, which smells of goats and apples. Animal and fruit scents are linked by the esters of fatty acids, which are chemically closely related so that they frequently occur together in the leaves of plants. The presence of caproic acid accounts for the fur-like smell.

BLACK CURRANT *Ribes nigrum*

A deciduous shrub without spines. The leaves 3–5 lobed, sticky beneath due to the presence of brown aromatic glands. The flowers, borne in pendulous racemes, are minute and greenish-yellow with stigma and anthers both ripe when the flowers open. They are visited by few insects, but this is not necessary as self-fertilisation readily takes place, the pollen falling on the recurved margin of the stigma. This is the only British species in which the leaves are aromatic or strong-smelling. The flowers are followed by jet-black globose fruits, ripe in July. In bloom May.

Habitat: Common throughout the British Isles except the far north of Scotland and Ireland, usually in damp deciduous woodlands and in hedgerows, also by slow-moving streams.

History: The thick juice of the berries concocted over a fire with sugar, formed a popular 'nightcap' during medieval times, while it was also taken for a sore throat and quinsy, hence its country name *Quinsy-berry.* A 'tea' was made by infusing the freshly gathered leaves in hot water. In France, a liqueur is made from the fruits, known as *Liqueur de cassis* while in certain countries a 'brandy' is distilled from the leaves.

MEZEREON FAMILY Thymelaeaceae

Mostly shrubs (few herbs) with mainly evergreen alternate or opposite exstipulate leaves, and with only a few exceptions inhabiting the warmer regions of the world, especially Africa. The plants have a tough inner bark, highly acrid, which may cause blistering of the skin. Flowers borne in condensed racemes; the perianth tubular, mostly 4-, rarely 5-cleft. Stamens 2–10, in 2 rows in perianth tube. Ovary superior; stigma undivided. Fruit, a 1-seeded nut or drupe.

DAPHNE Evergreen or deciduous shrubs with alternate, short-stalked leaves. Flowers borne in terminal or axillary racemes; perianth tubular with 4 spreading coloured sepals. Petals absent; stamens 8; filaments short. Ovary 1-celled. Fruit, a drupe.

Almost all species bear highly scented flowers in which indol is present but in association with methyl anthranilate to give a pleasing lightness, similar to Jasmine blossom. The essential oil contains benzyl acetate and like all such flowers where this occurs, they are coloured reddish-pink or purple-pink and are pollinated by butterflies. Those pollinated by moths are white, e.g. *Daphne alpina, D. blagayana,* and are most fragrant at night. In bloom January–May.

MEZEREON *Daphne mezereum* Pl. 11

A spreading deciduous shrub of extremely slow growth, a fully grown plant seldom exceeding 4–5 ft in height. The leaves lanceolate, stipulate. The flowers are purple-pink and appear before the leaves. The honey is secreted at the base of 6-mm long corolla-tube and is mostly visited by the Red Admiral (*Vanessa atalanta*) which is also attracted by the purple-pink flowers of the Buddleia. Honey-bees also act as pollinators, rubbing the proboscis against the two whorls of anthers near the end of the tube, then bringing it into contact with the stigma at a lower level before reaching the honey secreted at the bottom of the tube. The small flowers, pubescent on the outside, appear before the pale green leaves. The fruit is a red berry which is poisonous to animals but not to birds. In bloom January–May.

Habitat: It is still to be found in parts of Hampshire, Oxfordshire and Buckinghamshire; also E Yorkshire, usually in a chalk or limestone soil and in the partial shade of deciduous woodlands.

History: 'Mezereon too', wrote Cowper, 'though leafless, well attired and thick beset with blushing wreaths, investing every spray'. There is some doubt as to whether the plant is a true native of the British Isles for Turner did not mention it and Gerard referred to it growing in damp woodlands of eastern countries especially Poland, though it grew in his own garden and in most cottage gardens of his time. It may have been present naturally in very limited areas. Philip Miller, whose *Gardener's Dictionary* was the first comprehensive work of its type, was the first to mention finding it growing naturally in woods near Andover in Hampshire, and Gilbert White found it at nearby Selborne, which suggests that it may have been indigenous to that part only.

SPURGE-LAUREL   *D. laureola*   Pl. 11

A low-growing evergreen shrub with smooth erect stems at the end of which are the lanceolate laurel-like leaves, leathery and glossy, which form tufts. The tiny greenish-white flowers appear in axillary clusters as if encircled by the leaves and are most highly scented at night. Pollination is by night-flying Lepidoptera. The flowers are followed by jet-black oval berries which are poisonous to man and to animals, leaving a burning sensation in the throat if eaten. In bloom March–April.

Habitat: Present in dense woodland throughout England and S Wales but rare in Scotland and Ireland.

WILLOW-HERB FAMILY   Onagraceae

Annual or perennial herbs with alternate, opposite or whorled simple exstipulate leaves. Flowers borne solitary in leaf axils or in spikes or racemes. Calyx of 2, 4 or 5 sepals, petals, and stamens, with the latter often in 2 whorls of 8. Ovary 1–6 but mostly 4-chambered. Fruit, a 4-chambered capsule or a berry. Few have any degree of scent, *Fuchsia* of New Zealand and South America being pollinated by ruby-throated humming-birds, and *Epilobium* with its inconspicuous pale red flowers receiving few insect visitors and being adapted for self-pollination.

Oenothera, with its large white or pale yellow flowers and long tubes, has exceptional scent in the evening and is visited by nocturnal Lepidoptera.

OENOTHERA Perennial, annual or biennial herbs with scattered exstipulate leaves spirally arranged. The flowers borne solitary or in small axillary racemes. Sepals reflexed; petals broad, usually overlapping; stamens in 2 groups of 4 each. Calyx-tube constricted to form the receptacle in which the deep-seated honey is accessible only to the longest-tongued moths. The flowers open in the cool of the evening. The petals are held in the calyx, the ends of which meet over the corolla and clasp each other by a hook. As the temperature falls, the corolla swells, inflated by what would appear to be a gaseous fluid. The swelling continues until it bursts open the hooks, when the petals expand to cup shape and release the sweet scent. In bad weather when day temperatures are low, the flowers may open, and as those of *O. biennis* are coloured pale yellow rather than the usual white, hive-bees are attracted and ensure fertilisation when night-flying Lepidoptera are scarce. After fertilisation, the flowers quickly lose their scent and die. In bloom July–late September.

LESSER EVENING PRIMROSE

Oenothera biennis

A biennial growing up to 5 ft tall, with smooth lanceolate leaves and bearing cup-shaped flowers of palest yellow about 2 in. across. The sepals are bright green, the petals broader than long and shorter than the calyx-tube. The flowers are followed by a long 4-sided pubescent capsule. In bloom July–late September.

Habitat: Not common, but naturalised in sandy places near the coast and by railway embankments throughout the British Isles but not N Scotland. Readily seeds itself.

Lesser Evening Primrose

History: A plant of North American origin, seed of which was first sent from Virginia to Padua in 1619 and may have reached England a year or so later. Parkinson, the first to mention it, wrote of it in familiar terms, calling it the *Tree Primrose* (because of its height) of North America. The poet Barton wrote:

You, Evening Primroses, when day has fled,
Open your pallid flowers, by dews and moonlight fed.

FRAGRANT EVENING PRIMROSE *O. stricta*

Syn: *O. odorata*

A biennial growing 3–4 ft tall, pubescent and glandular. The lanceolate leaves, narrowing at the base, have curled margins and are toothed. The flowers, which measure more than 2 in. across, are deep yellow, turning orange after fertilisation, and emit a sweet, penetrating perfume as they open. In bloom July–September.

Habitat: It has become naturalised in parts of Devon and Cornwall, near river estuaries and banks and on wasteland close to the coast, but is rare elsewhere.

History: It was introduced into Britain in 1790 by Sir Joseph Banks who obtained from the surgeon of a merchant ship seed which had been collected on the coast of Patagonia, its sole native habitat.

CORNEL FAMILY Cornaceae

Trees or shrubs with opposite or alternate simple leaves, rarely exstipulate. Flowers small, borne in clusters or heads. Calyx-limb 4–5-toothed or lobed; petals none, or 4 or 5; stamens 4 or 5. Ovary inferior; fruit, a drupe with 1–4 stony seeds.

THELYCRANIA Deciduous trees or shrubs with opposite, entire leaves; exstipulate. Flowers hermaphrodite, borne in terminal cymes. Calyx-teeth small. Honey secreted in the fleshy disc surrounding the base of the style, which lies on the flat, exposed surface. The flowers are pale yellow or creamy-white with purple anthers and are visited by Diptera, which are attracted by the fetid smell and purple colouring of the anthers. Coleoptera also visit the flowers.

CORNEL, DOGWOOD *Thelycrania sanguinea*

Syn: *Cornus sanguinea*

A deciduous shrub growing 6–8 ft tall and almost as wide, with opposite acute leaves which turn reddish-bronze in autumn and small creamy-white flowers borne in corymbose cymes. Stamens 4, alternate with the petals. Anthers and stigma develop together and are level with each other so that both organs are touched by the smallest of the short-tongued insects which are attracted by the fetid smell. The flowers are followed by black fruits. In bloom June–July.

Habitat: Common throughout the British Isles, abundant in copses and hedgerows where the soil is of a calcareous nature, and producing numerous suckers. It is especially prevalent from Durham northwards to the Kingdom of Fife.

History: It takes its older botanical name from the Latin *cornu,* a horn, which denotes the hard, horny texture of its wood. This provided the material with which the ancients made their javelins. Virgil in *The Georgics* wrote:

The war from stubborn myrtle
   shafts receives;
From cornels, javelins.

In England it was known as the *Prickwood,* from its use in making skewers. 'Our common Dogwood', wrote Evelyn, 'is made use of for

Dogwood

cart timber, and rustic instruments; for mill-cogs, spokes, bobins, tooth picks and butcher's skewers'. Parkinson said that its fruit was so bitter that it should not even be given to dogs, hence its name of *Dogwood.* It received its botanical name *sanguinea* from its crimson leaves in autumn and the red colour of its young twiggy branches which seem to be ablaze when the winter sun shines upon them, making the plant a valuable addition to the shrubbery.

PARSLEY FAMILY   Umbelliferae

Mostly herbs of the northern temperate zone generally with stout, hollow stems and alternate, exstipulate leaves, their blades pinnately divided and sheathing at the base. Flowers small, usually white or cream, occasionally pink, borne in compound umbels, each flower being stalked. Calyx superior, 5-toothed; petals and stamens 5, alternating; stamens spreading further apart than in Cornaceae and with the honey even more accessible owing to the slightly convex disc. Also with the flowers rayed and uniting in almost a single plane, they are more readily visited by all the shortest-tongued insects including flies, wasps and beetles, with the latter confining their attentions to those species bearing white flowers. The scent of several of the umbellifers is unpleasant and usually attracts the insects which themselves smell unpleasant, e.g. *Anethum graveolens* which is visited by a species of *Prosopis* of similar smell. The genera which contain a poisonous sap usually emit a nauseating smell as a warning, e.g. Hemlock, all parts of which contain the poisonous alkaloid connia. In others, aromatic oils are present in the fruit, e.g. Caraway, Coriander, while the stems and roots of Angelica and Eryngo are sweet and aromatic. The 2 carpels are united into a 2-chambered ovary, crowned by the fleshy disc; the carpels usually adnate to a simple axis containing resinous ducts or furrows.

ASTRANTIA Perennial herbs with a creeping rootstock and leaves palmately divided. Flowers borne in a simple or compound umbel; bracts large and coloured; flowers male and hermaphrodite; calyx teeth longer than the notched petals. Fruit ovoid with toothed ridges. The flowers have a fetid smell and seem to be pollinated by bugs rather than by flies.

GREATER OR PINK MASTERWORT
Astrantia major

A glabrous perennial growing 2 ft tall with palmately 3–7 lobed leaves, whitish on the underside with bristle-like hairs. The lanceolate bracts are white below, pinkish-purple above, while the pinkish-white flowers are borne in a small rounded umbel, resembling Scabious. They have a sickly fetid smell like *Sorbus*. In bloom July–August.
Habitat: Naturalised in parts of Herefordshire and Worcestershire, close to the Welsh border; rare elsewhere. Usually present by the edge of woodlands or in hedgerows, enjoying partial shade.

Pink Masterwort

ERYNGIUM Erect spiny perennial herbs with fleshy wax-coated roots, and leaves reduced to spinous sheaths. Flowers borne in dense heads surrounded by spiny bracts which make it difficult for insects to pass over them. Calyx-teeth longer than petals, which are narrow, notched. Honey is secreted in a hollow disc and is accessible only to the longer-tongued insects, mainly wasps. The flowers are devoid of scent but the roots have a sweet aromatic smell.

SEA HOLLY
Eryngium maritimum
An erect herbaceous perennial with spiny blue-green leaves resembling those of Holly and with white veins. The ovoid flower-heads are sky-blue with spiny bracts, while the rhizomatous roots extend far into the ground like the Parsnip. They have a peculiar sweet smell and a pleasant sweet taste. The flowers are borne on 12–24-in. stems. In bloom July–August.

Habitat: Common around the coast of Britain especially Kent and Essex; growing in sand and shingle by the seashore.

History: In Elizabethan times and earlier, the roots were candied and were known as *eringoes* and were referred to by Shakespeare in *The Merry Wives of Windsor*, when Falstaff, wearing a buck's head, greets Mistress Page and Mistress Ford:

Sea Holly

Mistress Ford:	Sir John? Art thou there my deer? My male deer?
Falstaff:	My doe with the black scut? Let the sky rain potatoes; let it thunder to the tune of Green Sleeves; hail kissing-comfits; and snow eringoes . . .

Shortly after Shakespeare's death, Robert Buxton of Colchester achieved national fame for his eringoes, made from the roots of the plant which then grew in abundance around the Essex coast. The roots were also boiled and served with sauce, like Parsnips.

CHAEROPHYLLUM   Hairy herbs, annual or perennial with pinnate leaves and bearing white flowers in compound umbels; petals notched; bracts few or absent. Fruit oblong, with short beak.

SWEET CHERVIL   *Chaerophyllum sativum*
Syn: *Anthriscus cerefolium*
Possibly not a native plant but long naturalised in localised parts as a garden escape. Distinguished from *Anthriscus* in that it has only 3 bracteoles. A pubescent annual growing 1–2 ft tall, it has hollow stems covered with silky hairs between the nodes and 3-pinnate leaves,

Plate 11
Mezereon Family   1 Spurge-laurel   2 Mezereon

Plate 12
Parsley Family   1 Wild Angelica   2 Spignel-meu
3 Wild Celery   4 Fennel

pubescent on the underside and emitting the sweet scent of aniseed when bruised. The flowers are borne in lateral sessile umbels, the pedicles shorter than the bracteoles. They are followed by smooth fruits with a long beak. In bloom May–June.

Habitat: In hedgerows and by waysides, usually near villages which have remained relatively undisturbed since Tudor times.

History: Its handsome fern-like foliage, Gerard wrote, 'is deeply cut and jagged, of a very good and pleasant smell . . . which has caused us to call it Sweet Chervil'. During Gerard's time the leaves were used in salads to impart their pleasant aniseed scent and flavour. They are used in this way in Europe today and also for garnishing, as they have the appearance of pieces of old lace. The roots also have the aniseed smell, due to the presence of anisic aldehyde and were described by Parkinson as being 'of a sweet, pleasant and hot spicie taste'. They were candied, like eringoes and he described them 'of singular good use to warm and comfort a cold phlegmatick stomach'. The roots, however, are poisonous unless first boiled. When eaten with oil and vinegar Gerard said, they 'rejoiceth and comforteth the heart'. William Langland, the ploughman-poet of Malvern, wrote in *Piers Plowman*: 'chibolles (shallots) and shervelles and ripe cheries manye'.

GOLDEN CHERVIL *C. aureum*

A hairy erect perennial resembling Wild Chervil or Cow Parsley, *C. sylvestre*, the most common umbellifer. The stem is solid, grooved and swollen at the nodes. It may be purple-spotted. The 3-pinnate leaves are bright golden-green, hairy on the underside and when handled emit the sweet but peculiar smell of liquorice. The flowers are borne in large rayed umbels and are followed by oblong fruits, almost yellow in colour and prominently ribbed. In bloom June–July.

Habitat: Central Scotland, usually in open meadows. Rare.

MYRRHIS Pubescent perennial herbs. Leaves pinnate; umbels compound, many-rayed; bracts absent; bracteoles numerous. Flowers white; calyx-teeth absent, petals notched. Fruit oblong, beaked, with prominent ridges.

SWEET CICELY *Myrrhis odorata*

An erect pubescent perennial herb growing 3–4 ft tall, its stem hollow and furrowed, its pale green fragrant 3-pinnate leaves large and downy. The white flowers, male and hermaphrodite, are borne in terminal umbels. The bracteoles are white. The flowers are followed by large dark brown fruits, sharply ridged, which smell faintly of cloves. In bloom May–June.

Habitat: The most common umbellifer of N England, present in
woodlands and hedgerows from
Derbyshire and Lincolnshire, north
to C Scotland. Rare in S England
and S Ireland.

History: It takes its name from the
Greek *murrha*, myrrh, for the whole
plant emits a myrrh-like smell,
sometimes likened to aniseed, when
handled. The leaves were once
popular in salads and have a sugary
taste. They were also boiled as a
substitute for spinach. The roots, too,
may be boiled and served with white
sauce while the seeds, with their refreshingly aromatic scent of cloves,
were ground and mixed with wax to
rub over furniture to create a polish
and to leave behind a pleasing perfume.

Sweet Cicely

CORIANDRUM Glabrous annual herbs. Leaves pinnately decompound. Flowers pink or white, borne in compound umbels;
bracts absent. Calyx-teeth unequal; petals globose. Fruit round with
ridges low and indistinct, crowned by the calyx-teeth.

CORIANDER *Coriandrum sativum* Pl. 22

An annual, growing 6–12 in. tall with solid branching stems which are
ridged. The dark green leaves are 2-pinnate with the lower divided into
deeply cut segments. The umbels of 5–8 rays are without a general
involucre and the partial ones consist only of a few small bracts. The
flowers are mauve-pink or white with longer outer petals. The fruit is
palest yellow, of peppercorn size with no oil ducts on the outer surface
but with two on the inner face of each half. The whole plant possesses
the unpleasant smell of bugs and is pollinated by Coleoptera and
Diptera which obtain the readily accessible honey. The seed, too, when
unripe, has a similar smell but if allowed to become completely ripe,
the offensive odour is replaced by a sweet scent resembling Orange
blossom. This is intensified in the essential oil which in appearance also
resembles the essential oil of Orange blossom. The seeds yield about
0.5% of oil, the alcohol coriandrol being the principal constituent.
When treated with acetic anhydride, it gives rise to limonene whose
physical properties are identical with those of licarene, a dextrogyre of
linalol, present in the essential oil of lavender and of similar scent. The

nature of the chemical action producing this change of odour in the fruit is not as yet understood, but the longer the seeds are kept under dry conditions the more pronounced their scent and essential oil becomes. In bloom June–August.

Habitat: A rare casual to be found mostly in Essex and Kent where it was at one time grown commercially to provide seed for children's sweets.

History: The plant, which takes its name from the Greek *koris*, a bug, is native of the Levant and N Africa and possibly reached Britain with the Romans. In the *Book of Numbers* (ch. xi, verse 7) manna is likened unto Coriander seed.

SMYRNIUM   Erect glabrous annual or perennial plants; leaves 3-ternate with broad segments. Umbels compound, bracts and bracteoles usually absent; sepals minute or absent; petals yellow. Fruit ovoid, black.

ALEXANDERS   *Smyrnium olusatrum*

A strong-growing annual or biennial attaining a height of 4 ft, with smooth, hollow, furrowed stems and broad bright green glossy leaves; the leaflets toothed. The greenish-yellow flowers are borne in a large rounded umbel and are followed by broad black seeds with acute ridges, which when ripe have an aromatic and spicy smell. In bloom April–May.

Habitat: Long naturalised in the British Isles, it is widespread in hedgerows and on waste ground, usually in sight of the sea and growing in a calcareous soil. It is common in Kent, NE Yorkshire and extending northwards to Aberdeen; also in N Ireland.

History: It was at one time called Alexandrian Parsley, *Petroselinum alexandrinum*, for its home is that part of the N African coast around Alexandria. It may have reached Britain with the Romans for the leaves were used to impart their unique myrrh-like flavour to broths and stews and were eaten raw in salads. They were also used, like Fennel, to make a sauce to accompany fish. The young shoots and tops of the fleshy roots were blanched and boiled.

CONIUM   Tall glabrous annual or biennial herbs, native of N Europe and S Africa and containing the volatile poisonous alkaloid, conia, which has a smell resembling that of wet rat fur. Leaves pinnately compound; umbels compound; bracts and bracteoles small; sepals absent; petals white, with inflexed tip. Fruit ovoid, with 5 ridges.

HEMLOCK   *Conium maculatum*

A handsome branched biennial growing 3–5 ft tall with smooth furrowed hollow stems spotted with red and purple. The 2- or 3-pinnate leaves are deeply serrated and like the stems are also smooth, emitting a nauseating smell when handled. The white flowers are borne in terminal and axillary umbels and have short peduncles. They secrete much honey and are pollinated by Diptera and Coleoptera, which are drawn by the unpleasant stench of the whole plant. The large globular fruits have waved ridges. In bloom June–July.

Habitat: Widespread throughout the British Isles but mostly in the south and usually in damp ground, by streams and in deciduous woodlands and hedgerows; also on waste ground.

History: It was a draught of Hemlock which caused the death of Socrates, though in qualified hands it has valuable medicinal qualities. To the countryman the dried stems were known as *kecksies* and they were included, together with the living plant, among those obnoxious weeds mentioned by the Duke of Burgundy in Shakespeare's *Henry V* when describing the condition of the French countryside following the prolonged struggle with England:

> Her fallow leas
> The darnel, hemlock and rank fumitory,
> Doth root upon: while that the coulter rusts,
> That should deracinate such savagery:
> . . . and nothing teems,
> But hateful docks, rough thistles, kecksies, burrs,
> Losing both beauty and utility.

Again, in *King Lear* it is included with harmless weeds draped about the feeble old king who came:

> Crown'd with rank fumiter and furrow'd-weeds,
> With burdocks, hemlock, nettles, cuckoo-flowers.

During Shakespeare's time Hemlock was associated with witches' brews and Ben Jonson includes it among other poisonous plants of the English countryside in the witch's song in his *Masque of the Queen*:

> I ha' been plucking (plants among)
> Hemlock, henbane, adder's tongue,
> Nightshade, moonwort, leopard's bane.

APIUM Annual, biennial or perennial herbs, glabrous and with furrowed stems. Leaves pinnate or ternate. Umbels compound, axillary or ternate; bracts absent; bracteoles many or absent; sepals absent. Calyx-teeth, none; petals entire. Fruit, ovoid or oblong.

WILD CELERY, SMALLAGE *Apium graveolens* Pl. 12
A glabrous biennial herb growing 2–3 ft tall with a much-branched

furrowed stem. The shiny 3-pinnate leaves are divided into toothed leaflets and when handled release the familiar pungent smell of celery. The flowers are greenish-white and borne in terminal and axillary shortly-stalked umbels. They are followed by tiny black globular fruits. In bloom July–August.

Habitat: Widespread by rivers and streams, especially in S England, becoming rare farther north; also in sight of the sea, usually in calcareous soils.

History: The whole plant, including the seed (fruit) has a pungent smell. The seeds are used as a condiment and to flavour cheese, while an infusion of them has for long been regarded as a remedy for rheumatism. From the wild form the cultivated variety, *dulce*, was obtained (also the Turnip-rooted Celeriac) which was first described by Parkinson, who called it 'Sweet Parsley or sweet Smallage . . . resembling fennell'. He said that 'the first year it is planted it is sweet and pleasant . . . after it hath grown high, it hath a stronger taste (and smell) of smallage'. From Parkinson's notes it would seem that the cultivated form originated in Venice for he mentions having first seen it in the garden of the Venetian Ambassador in Bishopsgate, and describes how the Venetians used to prepare it.

PETROSELINUM Annual or biennial herbs. Leaves 3-pinnate, shiny, crimped. Flowers yellowish-white, borne in compound umbels; calyx-teeth absent, petals with small notch. Fruit globular with 5 ridges.

PARSLEY *Petroselinum crispum*
A bright green glabrous biennial growing 9–12 in. tall with solid stems. Leaves 3-pinnate, the lower much crisped or crimped and emitting a clean pungent smell when bruised. The yellowish flowers are borne in flat-topped umbels; bracts 2–3; bracteoles 5. Pollination is almost entirely by Diptera. In bloom June–August.

Habitat: Found in hedgerows and on old walls as a garden escape. It is also occasionally present on rocks and cliffs close to the sea, especially in S England.

History: The plant was described in the fourteenth-century work on gardening by Master John, though Turner, writing in 1548, said that he had seen it 'nowhere but in gardens', thus indicating that it had been introduced, possibly by the Romans or early in our history. Timbs recorded that Charlemagne, after eating cheese containing Parsley seed, so much enjoyed it that he had two cases of such cheeses sent each year to Aix-la-Chapelle. Shakespeare knew it well and in *The Taming of the Shrew* Biondello coarsely remarks that he knew 'a wench married in an afternoon as she went to the garden for parsley to stuff a rabbit.'

It is said that Henry VIII enjoyed it above all other accompaniments to meat or fish. Edmund Spenser wrote:

> Fat coleworts and comforting perseline
> Cold lettuce and refreshing rosemarine.

In Russell's *Boke of Nurture*, we are told that 'Quinces and Pear syrup with parsley roots (were) right to begin your meal' thus indicating that the aromatic roots were also used, and this is confirmed by Francis Bacon who in his *Sylva Sylvarum* said the roots were good to eat boiled, or raw if young. Gerard said that the roots or seed 'boiled in ale cast forth strong venom or poison'. The seed, pleasantly fragrant, is also used as a condiment. The Greeks made use of the leaves for bridesmaids' garlands and to give to victors at the Olympic Games. The root contains the chemical principle, apiin and a volatile oil, apiol, which was first obtained in Brittany in 1849 by Dr Joret. It acts as a sedative for nervous disorders of the head and spine.

CORN PARSLEY *P. segetum*

A biennial with a slender wiry erect stem 18 in. tall and with small pinnate leaves, serrate or lobed, brilliant green in colour and yellowing with age. The lower leaves quickly wither. The flowers are white and borne in irregular umbels with the petals notched. The ovoid fruits and the stem emit the smell of Parsley, though less pronounced. In bloom August–September.

Habitat: Present in cornfields and on waste ground, usually near the sea and most common in Kent and NE Yorkshire; in Dovedale and S Wales. Rare elsewhere.

SISON   Glabrous biennial herbs with pinnate leaves. Flowers white, borne in compound umbels; bracts and bracteoles several; calyx-teeth absent; petals notched. Fruit ovoid with 5 ridges.

STONE PARSLEY *Sison amomum*

A slender glabrous biennial growing 2–3 ft tall with solid wiry branched stems. The long petioled 2-pinnate leaves have narrow upper leaflets. When bruised they release an ammonia-like smell, resembling stale perspiration, which attracts dung-flies and midges for its pollination. The white flowers with their deeply notched petals are borne in terminal or axillary umbels. Bracts and bracteoles 3–4. In bloom July–August.

Habitat: Widespread in hedgerows and on roadsides throughout England and Wales, especially in calcareous soils.

CARUM   Glabrous perennial or biennial plants with slender hollow stems. Leaves pinnate. Flowers white or pink, borne in compound umbels. Bracts and bracteoles several or absent; calyx-teeth minute; petals notched. Fruit oblong with 5 slender ridges.

CARAWAY   *Carum carvi*   Pl. 22

A glabrous biennial with hollow stems 2 ft tall and a parsnip-like root running deep into the ground, which has an aromatic smell and a flavour similar to the Parsnip's. The 2-pinnate leaves are cut into linear lobes and resemble those of Parsley. They have the same culinary uses. The white flowers are borne in large irregular umbels with usually a single bract. The fruit (the carpels) is oblong with small ridges and has a powerful scent when mature and fully dry. In bloom June–July.

Habitat: A rare plant usually found on waste ground in heavy soils retentive of moisture, chiefly in Essex and East Anglia. Extremely rare in the north and west.

History: It takes its name from *Caria* in Asia Minor but is prominent over the whole of the northern hemisphere from Iceland to Siberia and south to the Mediterranean and the Middle East. It may well be native to the British Isles. The fruits are about ⅛-in. long with 5 low ridges and dark-brown channels between. Their essential oil consists of two portions, carvene ($C^{10}H^{16}$) and carvol ($C^{10}H^{14}O$), the former a hydrocarbon isomeric with turpentine, the latter an oxidised solid resembling camphor. The two bodies may be separated by fractional distillation. The value of the seed depends upon the amount of carvol present and the extent of its purity. It becomes pure upon removal of the carvene. The carvol of caraway is identical to that obtained from oil of dill and is often mixed with oil of lavender and bergamot for the manufacture of cheap perfumes and soaps. Ground caraway seeds may also be used to mix with other seeds to make sachets, for it releases its refreshing camphor smell over a long period. The essential oil is used for the flavouring of the liqueur, Kummel, while in Ireland the seed is baked in bread and cakes. In Shakespeare's time the seed was used to add flavour to apples. It was Mr Justice Shallow in *Henry IV*, *Part 2*, who said, 'you shall see my orchard where in an arbour we will eat a last year's Pippin of my own grafting, with a dish of caraways'. And from the accounts book of Sir Edward Dering (1626) there is a sum given for this conclusion to a meal – 'apples with caraways'. The distilled water from the seed (gripe water) is given to children to relieve 'wind' for which purpose the seed was served in Hall of Trinity College, Cambridge – with roast apples. All parts of the plant are fragrant and in the east an oil of inferior quality is distilled from the husks and is known as 'chaff oil'. Under commercial cultivation, the yield of seed from an acre of ground is about 1 ton.

OENANTHE Smooth aquatic perennials with tuberous roots. Stems erect, hollow and inflated. Leaves 1–3-pinnate; lower, lobed and ovate; upper, linear-lanceolate. Flowers white, borne in compound umbels; scented in *O. fistulosa*. Calyx-teeth acute; petals notched. Fruit globose, surrounded by cork-like ridges and crowned with erect styles.

That the plants usually grow in shallow water may be the reason for their scented flowers and the necessity to provide greater attraction for pollinating insects, as they are often found growing solitary or in small groups, away from others. Besides Diptera, their pollinators would seem to be Hymenoptera and Lepidoptera, attracted by the pink tinging of the white flowers and their scent.

WATER DROPWORT   *Oenanthe fistulosa*

A glabrous aquatic perennial growing 12–18 in. tall and forming roots from the lower stem-nodes. The root-leaves are 2-pinnate with flat leaflets and are often submerged; the upper are 1-pinnate with thread-like segments. The scented flowers are pinkish-white and appear in domed heads, making them more conspicuous. Bracts absent but bracteoles numerous. The fruit is angular and crowned with erect styles. In bloom July–September.

Habitat: Widespread but rare away from the Fenlands and Midlands, though present in low-lying ground in Kent and Cornwall. It is found in water meadows; in marshland; and in shallow water, often partially submerged.

History: It takes its name from two Greek words *oinos*, wine, and *anthos*, a flower, from the unusual wine-like scent, like the aroma of mature port which is most pronounced in *O. fistulosa*.

FOENICULUM Tall-growing glabrous biennial or perennial plants with 3–4-pinnate leaves and narrow leaflets with sheathing bases. Flowers yellow, borne in compound umbels. Bracts and bracteoles usually absent; also calyx-teeth. Petals entire, with inflexed tip. Fruit ovoid with conspicuous ridges, distinguished from *Anethum* in being compressed from side to side.

FENNEL   *Foeniculum vulgare*   Pl. 12

A stout erect perennial growing 4–5 ft tall, its 3–4-pinnate leaves divided into numerous narrow segments, the whole plant having a blue-green appearance. The yellow flowers are borne in large terminal umbels and are followed by narrow ovoid fruits blunt at the ends and with 8 longitudinal ribs. All parts of the plant except the flowers are scented. In bloom July–September.

Habitat: Present on waste ground and especially on cliff tops throughout the British Isles but in greatest abundance around the coast of the southern part of England and Ireland; also Wales.

History: It is a plant steeped in the history of these islands and may have been introduced by the Romans who named it *Foeniculum* because of its hay-like smell, though anethol or anise camphor forms the principal constituent of its essential oil. It is present also in oil of anise, *Pimpinella anisum*, with which it is often mixed and which yields a camphor-like smell. Oil of fennel is used to perfume soaps, and the dried leaves are used in sachet powders and in pot-pourris. The seed was at one time used (as a substitute for juniper berries) to flavour gin and was supposed to allay the pangs of hunger, hence it was in great demand by the poor of medieval times. In *Piers Ploughman*, a priest asks a woman for 'a farthingworth of fennel seed for fastyng days'. The accounts of Edward I (for 1281) show that eight and a half pounds of seed were bought for the Royal household for only one month's supply. From the leaves, a sauce was made to serve with mackerel, which is referred to by Shakespeare in *Henry VI, Part 2*, when Falstaff mentions the knight Poins' liking for 'conger and fennel' (p. 21).

The poet Longfellow in 'The Goblet of Life' best described the virtues of Fennel:

> Above the lowly plants it towers
> The fennel with its yellow flowers,
> And in an earlier age than ours,
> Was gifted with the wondrous powers
> Lost vision to restore.

He goes on to describe the qualities of strength and fearlessness bestowed on soldiers and gladiators by Fennel (p. 21).

SILAUM   Glabrous perennial herbs with solid stems. Leaves 1–3-pinnate; umbels compound; bracts 1, 2 or absent; bracteoles numerous. Flowers yellow; petals broad. Fruit ovoid with slender waved ridges.

PEPPER SAXIFRAGE   *Silaum silaus*
Syn: *Silaus flavescens*

A glabrous perennial with stiff, solid angular stems and growing 1–2 ft tall. The 3-pinnate leaves with their linear opposite leaflets release a fetid smell when handled and if consumed by cattle give an unpleasant smell and taste to their milk. The pale yellow flowers are borne in terminal umbels and are followed by ovoid fruits with waved ridges. In bloom June–September.

Habitat: Widespread throughout the southern half of England but rare in the north and absent from Ireland. Found in damp meadows, in hedgerows and on waste ground, usually growing in heavy soil.

MEUM   Glabrous perennial herbs growing 12–18 in. tall with pinnately decompound leaves, the segments composed of whorls of needle-like leaflets. Umbels compound; bracts absent; bracteoles 3–8. Flowers pinkish-white; petals inflexed. Fruit ovoid, ridges slender.

SPIGNEL-MEU   *Meum athamanticum*   Pl. 12

A glabrous perennial with hollow stems, all parts of which are aromatic, especially the parsnip-like rootstock. The 2-pinnate leaves are divided into segments composed of numerous needle-like leaflets which, when crushed, release an aromatic smell like turpentine. The pinkish-white flowers are borne in compound umbels and have the appearance of *Gypsophila*, with the petals inflexed and narrowing at both ends. They are followed by aromatic ovoid fruits with slender ridges. In bloom May–June.

Habitat: A plant of Scotland and N England, present on mountainous slopes of pastures; also N Wales.

History: At one time the roots were used as food in the Highlands of Scotland and the leaves were boiled to serve with meats. It is known to Northcountrymen as *Bald*, a name derived from *Balder*, a handsome god of the Norsemen to whom the plant is dedicated.

SELINUM   Glabrous perennial herbs with stout rootstock. Leaves 2-pinnate; leaflets pinnatifid. Umbels compound, many-rayed; bracts absent; bracteoles numerous. Flowers white; petals notched. Fruit ovoid, 10-winged.

CAMBRIDGE PARSLEY, FALSE MILK PARSLEY
Selinum carvifolia

An erect glabrous perennial with solid angled stems. The 2-pinnate leaves are dark green with pointed serrated leaflets and when handled emit a parsley-like smell. The white flowers are borne in large terminal umbels, 20-rayed and

Cambridge Parsley

flat-topped. They are followed by 10-winged ovoid fruits. In bloom August–September.

Habitat: Present only in a few fenland meadows around Cambridge and Ely and in S Lincolnshire.

ANGELICA   Perennial herbs confined to the northern hemisphere and New Zealand. Stems hollow. Leaves 2-pinnate with large segments. Umbels compound, many-rayed; bracts few or absent; bracteoles numerous. Flowers greenish-white or pinkish-white; petals incurved. Fruit ovoid with 2 wings.

WILD ANGELICA   *Angelica sylvestris*   Pl. 12

A tall-growing perennial with hollow stems often of purple colouring and slightly downy. Leaves 2-pinnate with the broad leaflets stalked and acutely serrated. The pinkish-white flowers are borne in large many-rayed umbels. Bracts and bracteoles few. Petals incurving; calyx-teeth absent. Fruits flat with 4 broad ridges. The plant has a slight musky smell about it which, however, is not nearly so pronounced as in *A. archangelica*, a naturalised garden escape. In bloom July–September.

Habitat: Widespread throughout the British Isles and common in damp woodlands and hedgerows.

GARDEN ANGELICA   *A. archangelica*

Syn: *A. officinalis*

An introduction, differing from the Wild Angelica in having minute sepals, while the stems are free of any purple colouring and the flowers are greenish-white. A perennial, it grows 4–6 ft tall with leaves 2–3 ft across with decurrent segments. The whole plant has a powerful musky scent, all parts being aromatic. In bloom August–September.

Habitat: A garden escape, it has become naturalised along river-banks, in marshland and moist hedgerows throughout the British Isles, but is rare; it grows in profusion, however, in Greenland, Iceland, Scandinavia and Denmark.

History: It was called *Angelica* because it is supposed to be of heavenly origin, while the ancient people of the north believed that the plant possessed many valuable qualities. Parkinson wrote that 'the whole plant, leaf, root and seed is of an excellent comfortable scent, savour and taste'. The plant was mentioned by Philippe de Comines in his account of the battle of Morat, when the leaves were used (with those of other herbs) to cure wounds made by an arquebus. The stems, stewed with rhubarb, impart a delicious musky smell and when boiled and candied in syrup of sugar will become a brilliant green and make a pleasant sweetmeat. Turner wrote that in his day candying was done abroad at Danske where the plant grows wild, 'for a friend in London called Master Alleyne, a merchant, who hath ventured over to Danske sent me a little vessel of these, well condited with honey; very excellent good!'

An aromatic oil is obtained both from the root and from the seed, the latter yielding 1.15% of oil which has a powerful musky smell and is used to flavour Vermouth and Chartreuse.

The plant contains free angelic acid which, in the root, is found mixed with a little valerianic acid. Indian Sumbul root, of the same plant Order, is also found to contain angelic acid and emits the same smell of musk. It also contains about 10% of balsamic resin which, in water, releases a similar musky odour. The musk-scented angelic acid is also present in oil of chamomile.

In his *Calendar for Gardening* for July, written in 1661, Stevenson made the pleasing suggestion that one should perfume one's house by burning Angelica seeds over a fire (p. 20).

BIRTHWORT FAMILY Aristolochiaceae

Herbs or shrubs with twining lianes. Leaves alternate, simple, exstipulate. Flowers solitary, borne in racemes or clusters; perianth whorled, attached to ovary below; tubular above. Stamens 6–12; ovary 3–6-chambered; style, 1; stigmas equal to chambers of ovary. Fruit, a 3–6-chambered capsule. Of the 2 British genera, *Asarum* may be considered an early and incomplete stage in the prison development of *Aristolochia*. *Asarum* is almost devoid of smell. It is visited by larger flies and is also self-pollinating. *Aristolochia* has an evil, fetid smell and is visited by dung-midges.

ARISTOLOCHIA Rhizomatous herbs with twining lianes. Cauline leaves stalked, axillary. Flowers with tubular perianth, swollen at base and lined with hairs; limb 3-lobed. Stamens 6, inserted on style. Fruit, a capsule.

BIRTHWORT *Aristolochia clematitis*

A glabrous perennial growing 2 ft tall, with woody rhizomatous roots. Stems erect, unbranched with large cordate leaves. The flowers are borne in groups of 4–8 and are yellowish-green with an evil, fetid smell. The erect corolla-tube is lined with reflexed hairs. Midges creep down past these into the swollen base of the tube but are at this stage prevented from leaving. The stigma is mature when the smell is at its peak, and is touched by the midges carrying pollen from other flowers, while the anthers remain closed. As the stigma withers, the anthers open and the flower-tube begins to bend downwards. The hairs also wither to release the midges which, covered with pollen, fly away to fertilise other flowers. The flowers immediately lose their evil smell after fertilisation. In bloom July–September.

Habitat: Naturalised in several parts of England, especially north from Sussex, into Oxfordshire, Staffordshire and Derbyshire. Rare in Scotland; absent from Ireland. Found on waste ground and on ruined houses and old walls.

Birthwort

History: The juice from the stems was once used by midwives to allay the pangs of childbirth, while that of certain South American species is used as an antidote to snakebite.

SPURGE FAMILY   Euphorbiaceae

Distributed throughout the tropical and temperate regions of the world; some cactus-like, some heath-like, containing a poisonous milky juice. Leaves opposite or alternate, simple, with deciduous stipules. Flowers small, borne cymose or racemose, several males grouped around a female flower (males and females on separate plants); often with bracts or a cup-shaped involucre with conspicuous yellow glands arranged about the margin, which in certain South American species secrete an aromatic resin. Perianth absent or 3-merous; stamens 4–20; ovary 2–3-celled. Fruit, a capsule, rarely a drupe.

MERCURIALIS   Annual or perennial herbs, creeping or branched, with milky juice present. Leaves opposite, stalked, serrate. Flowers borne in axillary spikes; dioecious; perianth 3-merous; stamens 8–20; ovary 2-chambered; styles 2; ovules 2. Fruit, a capsule.

DOG'S MERCURY *Mercurialis perennis*

A hairy perennial with a creeping rhizomatous rootstock and an un-branched stem 12 in. tall. The dark green serrate lanceolate leaves are rough and hairy. The small greenish-yellow flowers are borne in spikes or racemes from the axils of the leaves. The male and female flowers appear on separate plants. The whole plant has a fetid smell for the leaves contain trimethylamine and when bruised give off the un-pleasant smell of decaying fish, a smell which in the words of Culpeper is 'somewhat strong and virulent'. It is pollinated by midges. In bloom March–May.

Habitat: In beechwoods and hedgerows throughout the British Isles. Rare in Ireland.

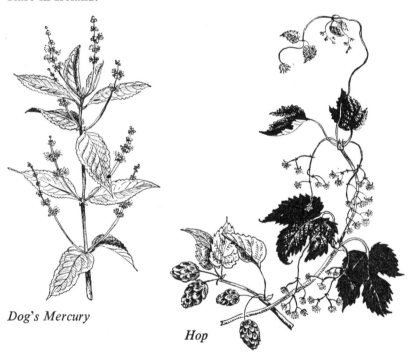

Dog's Mercury

Hop

HOP FAMILY Cannabinaceae

Herbs, sometimes climbing but stipules free. Leaves opposite, palmately lobed or entire. Flowers borne axillary in glandular inflorescence; females sessile; perianth entire with central style; stigmas 2. Fruit, an achene; seed with endosperm.

HUMULUS Climbing perennial hairy herbs with angled stems. Leaves opposite, palmately lobed, serrated. Male flowers borne on a branched inflorescence; females on a different plant, in axils of bracts to form a cone-like spike. Bracts persistent. Though wind-pollinated, the flowers have a slight perfume, while the leaves of *H. lupulus* are refreshingly aromatic.

HOP *Humulus lupulus*

A hairy perennial, climbing to a height of 15 ft. The 3–5-lobed leaves, cordate at the base, are coarsely serrated with a petiole as long as the blade. The flowers are greenish-yellow, the males borne in a catkin-like inflorescence, the females in the axils of the bracts to form a cylindrical spike, like 'scaled Pineapples', wrote Gerard, the reference being to pine-cones, 'of a whitish colour, tending to yellow and strong of smell'. The bracteoles bear glandular hairs yielding a golden dust, lupulin, an important ingredient in the brewing of beer. In bloom July–August.

Habitat: In hedgerows and woodlands throughout England. Rare in Scotland and in Ireland. May be planted to cover a trellis or bower.

History: It takes its name from the Anglo-Saxon *hoppian*, to climb, and if gathered fresh the flowers yield a fragrant otto or attar upon distillation, which is dark green in colour. 'Hops' (the seed capsules) were at one time used to stuff pillows, for when they are completely dry they emit a pleasing fragrance and have a slight narcotic effect, encouraging sleep. George III is reputed to have always slept on a hop pillow. Hops were not used for brewing in England until early in the sixteenth century. Before then, ale was made from malt and clarified by Ground Ivy (Ale-hoof).

WALNUT FAMILY Juglandaceae

Trees with alternate, pinnate leaves and flowers monoecious, borne in axils of bracts; males in drooping catkins; females solitary or in terminal spikes; bracteoles 2; perianth absent or minute. Ovary of 2 carpels. Style short, stigmas 2. Fruit, a drupe without endosperm.

JUGLANS Deciduous trees with alternate, pinnate leaves, pleasantly aromatic. Male flowers borne in drooping catkins; females in erect terminal spikes. Wind-pollinated and without scent. Bracteoles present in both sexes, united with ovary but not persisting in fruit with its hard endocarp (shell).

WALNUT *Juglans regia*

A tree of large proportions with smooth grey bark and glabrous grey twigs. The pubescent leaves are divided into 9 serrate leaflets, grey-green in colour tinted with red, which when bruised release a sweet, resinous smell similar to that of the Balsam Poplar which keeps away flies during summer. The fruit (plum) is covered with a green husk which stains the fingers black when removed to reveal a wrinkled, bony shell containing the nut. The shell is opened by two valves. Gerard said that 'the green and tender nuts boiled in sugar and eaten as "suckarde" are a most pleasant and delectable sweetmeat to comfort the stomach'.

Habitat: Usually planted as a specimen tree in a courtyard or at a focal point of the landscape because of its enormous spread. Naturalised in parts of the southern half of England, where in the warmer climate it grows best.

History: Exactly when the Walnut first came to be grown in England is not known for certain, but from Parkinson's notes it would appear to have been long established, and in his *Diary* for 21st September, 1665, written at the height of the Great Plague, Pepys wrote, 'To Nonsuch (Henry VIII's great palace), to the Exchequer by appointment, and walked up and down the house and park . . . a great walk of an elme and a walnutt set one after another in order . . .', thus indicating that the plantings would most likely have been made during Henry's reign, for at the time of Pepys' visit they were well established.

The Romans named the tree after Jove, *Juglans* meaning *Jove's nut*. It is called *Juglantis vel nux* in Aelfric's *Vocabulary*. By the fourteenth century it had received its names *Walnutte* and *Walshenut*. Lyte (1578) described the tree as being called 'the Walnut and Walshe-nut tree' in English. Gerard confirmed this.

The tree was possibly introduced into Britain by the Romans, for it had reached Italy from Persia at an early date. The tree was taken as their crest by the family of the poet Edmond Waller after the Battle of Agincourt, for during the battle the head of the family had achieved honour by taking the Duke of Orleans prisoner. The name *Waller* is said to be derived from the word *walnut*. Evelyn, writing in 1658, forty-two years after the death of Shakespeare, mentioned mature plantations at Leatherhead in Surrey, while London has described how a traveller went from Florence to Geneva living entirely on walnuts.

Shakespeare would almost certainly have seen the Walnut trees growing at Nonsuch, Henry VIII's palace at Ewell in Surrey, for its building commenced in 1538 and, as Pepys tell us, the Walnuts were mature by the middle of the following century. Yet Shakespeare made no reference to the tree itself, only to the nutshell.

In *The Merry Wives of Windsor* (and Walnuts must surely have been growing in the Great Park at the time) in a room in Mr Ford's house where Falstaff is present with Mr Page and the wives of the two men, Ford says:

> . . . let them say of me, As jealous
> as Ford, that searched a hollow walnut
> for his wife's leman.

BOG MYRTLE FAMILY Myricaceae

Aromatic trees or shrubs, mostly inhabiting North America. Leaves alternate, simple, with resinous glands. Flowers dioecious, catkin-like; stamens 2–6; ovary 1-chambered; styles 2. Fruit, a 1-seeded drupe, covered in resinous wax.

MYRICA   As above, with stamens 4–8. Flowers borne solitary in axils of bracts. Perianth absent; filaments free or united below. Females with 2 or more bracteoles; ovary 1-celled; style short; stigmas 2. Fruit a drupe.

SWEET GALE, BOG MYRTLE *Myrica gale*

A deciduous perennial shrub growing 3–4 ft tall and increasing by underground suckers. The oval grey-green leaves, pubescent on the underside, are serrated near the apex and are short-stalked. The smooth reddish-brown twigs are covered, like the leaves, in aromatic glands and release a refreshing resinous scent when handled. The brownish-green flowers are borne in catkins on the previous year's shoots and on separate plants. They are wind-pollinated and are *Bog Myrtle* scentless. The small berries (drupes) with 2 wings, are covered in resinous wax which, when treated with hot water and skimmed, yield 'myrtle' wax which emits a powerful balsamic perfume. It was at one time used to polish furniture and to make candles which diffuse a resinous odour when lighted and burn with a clear white flame. In bloom April–June.

Habitat: Present in bogs and heathlands, chiefly in Somerset and Devon; also in Suffolk. Absent from the Midland counties and the north.

BIRCH FAMILY Betulaceae

Evergreen or deciduous trees or shrubs. Leaves simple, alternate. Stipules membranous. Flowers monoecious, males in drooping catkins; females in erect catkins with 3-lobed scales. Perianth absent; stamens 2–20; ovary 2–3-celled; styles 2–3, free. Fruit, a small dry, winged, 1–2-seeded nutlet surrounded by involucre. Seeds large and fleshy.

BETULA　Deciduous trees or shrubs with simple, alternate leaves. Flowers borne in catkins with deciduous scales; petals absent; ovary 2-celled; styles 2, free. Fruit, a small, winged, 1-seeded nutlet. Flowers wind-pollinated and scentless but leaves of *B. alba* var. *odorata* are scented. The bark yields a fragrant oil (tar) which when used to treat leather gives the characteristic smell known as *Russian leather*.

WHITE OR SILVER BIRCH　*Betula alba*
Syn: *B. verrucosa, B. pendula*
A deciduous tree growing 50–60 ft tall with smooth white bark peeling transversely; the twigs shining and brown. The irregularly serrated leaves, held on long stalks and truncate at the base, have a resinous scent in the variety or subspecies *odorata*. With this form, the brown twigs are covered in resinous spots. The deciduous catkins are about 2 in. long and are followed by 2-winged fruits. In bloom April–May.
Habitat: B. alba is widespread throughout the British Isles, especially on hillsides and in well-drained gravelly soil in N England and Scotland, where it seeds to form pure woodlands.
History: A nourishing beer was made from its bark and by destructive distillation an aromatic oil was obtained, dark brown in colour and with a musky odour used in Russia to cure leather to which it imparted its unusual perfume. The oil was also used in medicated soaps for treating eczema. That the Silver Birch gives of its fragrance to those who seek it was foretold by the Scottish poet, William Hamilton, in 'The Braes of Yarrow':

> Flows Yarrow sweet? as sweet flows Tweed:
> As green its grass, its gowan as yellow;
> As sweet smells on its braes the birk,
> The apple frae its rock as mellow.

The Sweet Birch, *Betula lenta*, of North America yields from its inner bark an aromatic oil smelling of wintergreen, while an agreeable 'tea' is made from its leaves. *B. lenta* oil is superior to that of wintergreen, *Gaultheria procumbens*, for the latter gives an oil which has the irritating smell of black pepper. *B. lenta* is a most attractive garden tree with heart-shaped leaves which release their aromatic smell when pressed.

BEECH FAMILY　Fagaceae

Large trees or shrubs with simple, alternate, stipulate leaves. Flowers staminate, males arranged in a globose catkin; females in small groups at base of male inflorescence. In males, perianth 4–6-lobed; stamens twice perianth lobes; in females, perianth 4–6-lobed; ovary 3–6-celled; styles 3 or 6. Fruit, a single-seeded nut without endosperm.

CASTANEA Deciduous trees, buds scaly, leaves lanceolate and serrate. Male flowers borne in erect catkins; stamens 8–20. Females usually in threes, within a prickly cupule; stigmas 6; ovary 5–8-chambered. All but a single ovary abort so that each bears a single nut. Nut 1-chambered; 1–3 seeded. Unlike Beech and Oak of the same family, which are wind-pollinated with scentless flowers borne in drooping catkins, *Castanea* is visited by most forms of insect, especially flies. It bears flowers in upright catkins with a powerful scent like that of Hawthorn and containing trimethylamine sweetened with anisic aldehyde.

SWEET OR SPANISH CHESTNUT *Castanea sativa*

A large deciduous tree with deeply furrowed dark brown bark and greenish-brown twigs, somewhat pubescent. The leaves are oblong-lanceolate and deeply serrated, glabrous above, pubescent on the underside; they turn bright yellow in autumn before they fall. The pale yellow flowers have conspicuous white anthers; the females are followed by shining brown fruits (chestnuts) enclosed in a brilliant green cupule covered in bristly spines. The colour of the nuts has given the distinctive colour 'chestnut-brown' to our language. In bloom July.

Habitat: Widespread in deciduous woodlands in S and SE England but rare in the north, for it prefers warmer climatic conditions than the scentless flowered Horse Chestnut, *Aesculus hippocastanum.*

History: Native of SE Mediterranean, the tree takes its name from Kastanum, a town in Thessaly where the trees were abundant. They were probably introduced into Britain by the Romans who made use of the fruits (nuts) for food. Chaucer was familiar with the 'chesten' tree and by Stephen's reign the Great Chestnut of Tortworth had attained large dimensions. That chestnuts were considered to be a valuable food since earliest times is borne out by a reference to their being provided for Henry I I's queen, Eleanor of Aquitaine, in 1160. In the Court Accounts for that year is an item for 3s. for 'castanear' (chestnuts), sent to the Queen while she was visiting Salisbury from the Palace of Wood-stock, 60 miles away, probably to eat on the way.

In Chaucer's translation of *The Romance of the Rose*, chestnuts are included with other fruits enjoyed at the time:

> Medlers, ploums, peres, chesteynis,
> Cherries of which many one fayre is.

From earliest times roasted chestnuts have been sold in the streets of London during wintertime. Chestnuts are also ground to use as flour and will impart their earthy smell and flavour to bread and chestnut-cakes. The whole nuts are also boiled and crystallised to be eaten as *marrons glacés*, a readily digestible food for invalids. The nuts are ripe in October, when they fall from the tree still enclosed in the spiny burr-like container (cupule).

WILLOW FAMILY   Salicaceae

Deciduous trees or shrubs. Leaves simple, alternate, stipulate. Flowers dioecious, borne in catkins, each flower subtended by a bract. Bracteoles and perianth absent, replaced by a cup-like disc in *Populus*. Stamens 2–30; ovary 1-chambered; styles 2; ovules many. Fruit a 2–4-valved capsule; seeds covered with silky hairs.

POPULUS Deciduous trees with heart-shaped leaves lobed or coarsely toothed and held on long petioles. Flowers borne in drooping catkins, appearing before the leaves. Buds with outer scales and flowers with cupped disc. Stamens 4–30, not united; stigmas 2–4-lobed; anthers purple-red. Pollination is by wind, though in several species the unfolding buds release a balsamic scent which perfumes the air around.

BLACK POPLAR   *Populus nigra*
A large deciduous tree with greyish-black bark (hence its name), which is deeply furrowed. The branches are widely spreading, the twigs covered in grey down when young. The reddish-brown buds are sticky with resin and emit a slight balsamic scent upon unfolding to release the rhomboid serrate leaves. Male catkins with 8–20 stamens; females with 2 stigmas. In bloom March–April.
Habitat: Rare in S England where it is usually found by river-banks and in damp woodlands. Trees are known to have lived 600 years. The form *betulifolia*, the Manchester Poplar, is a beautiful tree for a town garden.
History: The variety *italica* is better known. This is the Lombardy Poplar, which forms a tall, narrow head and is a tree of rapid growth. The vessels in which grapes are carried from the vineyards of Lombardy are made entirely from this wood and last from 40 to 50 years.
   Of the native Black Poplar, Culpeper wrote in his *Herbal* (1653): 'The clammy buds thereof, before they spread into leaves are gathered to make Unguentum Populneum . . . and are somewhat small, sweet but strong (smelling)'. 'The ointment', wrote Matthiolus, 'is singularly good for inflammations in any part of the body'. Phillips wrote that if the resin is removed from the buds by spirits of wine, the substance obtained smells like storax.

BALSAM POPLAR, BALM OF GILEAD
P. balsamifera
Syn: *P. tacamahacca, P. gileadensis*
A large tree with spreading branches and brown, pubescent twigs. The

large buds are thickly covered in balsamic resin and unfold to release heart-shaped leaves tapering to a point, pubescent on the underside and with long petioles. The female flowers, borne in pendulous catkins, have a pink stigma. In bloom March–April.

Habitat: Occasionally found naturalised in damp woodlands as garden escapes, and increasing by means of suckers.

History: The tree reached Britain from North America in 1692 and came to be widely planted in gardens because of its extreme hardiness and ability to withstand clipping. The buds release a heavy balsam-like scent as they unfold and the leaves retain this all their life. If held over a low flame, the buds will release their scent throughout a large house, but the resin is so sticky that it has to be removed from the hands with turpentine. In North America and Siberia a medicated wine is made from the buds, while game-birds which feed on the buds during winter acquire a unique flavour much appreciated by epicures.

SALIX Small deciduous trees or shrubs inhabiting moist situations. Leaves longer than broad with prominent stipules. Bud enclosed by single scale. Leaves simple, alternate. Flowers borne in erect or horizontal catkins; males yellow and silky, stamens usually 2, occasionally 1–12; females green, on different plant; stigmas 2. Flowers appear before the leaves. They have crimson or purple anthers and, being erect, have a slight unpleasant smell to attract Diptera for their pollination. Conspicuous by their appearance before the leaves, the flowers possess an abundance of honey and pollen and are also visited by all species of Hymenoptera. The foliage of *S. pentandra* is pleasantly scented when young, and from the fresh leaves an oil is obtained by fermentation and distillation and by treating the distillate with ether. The oil is yellow with an agreeable aromatic odour like that of castor – the secretion of the castor beaver which feeds on the plants. It is a scent similar to that of the fresh leaves. Fruit, a 2-valved capsule.

BAY WILLOW *Salix pentandra*
Small deciduous tree with dark brown bark and glossy brown glabrous twigs. The thick elliptic-lanceolate leaves 4 in. long, are about 4 times as long as they are broad and are dark green above, pale green on the underside. They are fragrant, especially when unfolding, and if removed at this time retain their scent when dry and may be used in pot-pourris. The scent is released from 2–3 pairs of glands situated at the end of the petiole. The flowers are borne in cylindrical catkins after the leaves expand, as this is the latest Willow to bloom and the only native species with 5 stamens. The anthers are golden-yellow. The female flower has 3 nectaries. In bloom June.

Habitat: Fenlands, by the side of ditches and dykes and damp woodlands from north of the Thames through the Midland counties to N Scotland. Rare south of the Thames and in Ireland.

History: The plant takes its botanical name from two Celtic words *sal* and *lis*, meaning near water, though *S. pentandra* is one of the few Willows tolerant of a dry soil. *Willow* is derived from the Anglo-Saxon *withig*, meaning pliant, for the twigs had all manner of uses including basket-making (Osiers), the making of hurdles and ropes, and thatching, while the bark was spun into thread for clothes. The bark contains the bitter principle salicine which was used to allay feverish colds until the introduction of quinine.

HEATH FAMILY Ericaceae

Shrubs with evergreen, opposite, simple leaves, common in South Africa, North America and Europe, few of which have scented attractions. Flowers hermaphrodite, calyx 4–5-cleft, persistent; corolla hypogynous. Stamens equal in number to corolla segments or twice as many; free from corolla. Anthers with 2 appendages. Carpels united to form a 4–5-celled ovary. Fruit, a capsule or berry. Pollination mostly by Hymenoptera; *Calluna* (Ling), wind-pollinated; without scent.

GAULTHERIA Evergreen shrubs, the source of oil of wintergreen, hence some confusion with the closely related *Pyrola* and *Moneses* which are known as the Wintergreens because of their evergreen foliage. Leaves alternate, broadly acute, serrate. Flowers borne in terminal panicles; stamens 10. Fruit, a berry-like capsule, usually edible.

AROMATIC WINTERGREEN *Gaultheria procumbens*
An evergreen shrub which increases by underground runners from which the stems rise. The obovate leaves are serrated with short petioles and the older leaves take on brilliant red tints in autumn. Oil of wintergreen is distilled from the leaves which release a sharp, pungent smell when bruised. The pure white flowers, borne on red stems, also release a similar scent and are followed by large edible crimson fruits which persist through winter. In bloom May–June.

Habitat: Perhaps a garden escape but present in pinewoods in N Scotland, though rare.

History: Introduced from Canada where at one time the leaves were used to make a delicious drink known as *Mountain Tea*. The Gaultherias are invaluable for providing ground cover, for they flourish in a peaty oil and in the shade of woodlands. They may accompany Rhododendrons and other plants enjoying similar conditions and will suppress weeds and provide colour throughout the year. They also give ground-cover to game-birds.

WINTERGREEN FAMILY Pyrolaceae

Evergreen perennial herbs with rhizomatous rootstock. Stems short, unbranched. Leaves alternate or opposite, simple, smooth. Flowers hermaphrodite, borne in bracteate raceme or cyme, drooping to one side. Sepals 5, persistent; petals 5, free; stamens 10. Fruit, a 5-chambered capsule.

MONESES Differs from *Pyrola* in that the leaves are opposite; the flowers solitary; anthers tubular. Nectar absent. Pollination mostly by night-flying Lepidoptera.

ONE-FLOWERED WINTERGREEN *Moneses uniflora*
An evergreen perennial growing 4–6 in. high with pale green, serrate, radical leaves. It bears a solitary, drooping, long-stalked white flower nearly 1 in. across, possessing a rich, sweet scent which is especially pronounced at night. In bloom June–July.
Habitat: Native of the Arctic regions and of N Scotland and the Orkney Islands, where it is to be found in pinewoods.

PRIMROSE FAMILY Primulaceae

Herbaceous perennials, occasionally annuals, inhabiting northern temperate regions. Leaves opposite or alternate, exstipulate. Flowers borne in scapes without bracts; calyx 5-cleft; corolla present (absent in *Glaux*), with as many lobes as clefts in calyx. Stamens equal in number to corolla-lobes and opposite to them. Ovary 1-chambered. Style single; stigma capitate. Fruit a capsule; seeds usually many and small.

The flowers are mostly yellow, with two distinct forms, a long-styled and a short-styled, and were the subject of a series of interesting experiments by Darwin. He discovered that pollen from one form applied to the other resulted in yields half as great again as those obtained from illegitimate crossings. Pollination is almost entirely by Hymenoptera which in most instances are able to reach the nectar secreted at the base of the long tube. In the case of the Primrose, *Primula vulgaris*, the flowers bloom when few bees are about and fertilisation is rare. Scent is present in most members of the *Primula* family and in the pink-flowered *Cyclamen hederifolium* which is mostly pollinated by butterflies.

PRIMULA Perennial herbs with rhizomatous rootstock, each joint terminating in an inflorescence. Leaves radical. Flowers borne in short-

or long-stalked umbels, rarely solitary. Calyx 5-toothed; corolla salver- or funnel-shaped, 5-lobed with long slender tube. Stigma protruding beyond the end of the tube or inserted halfway down. Capsule 5-valved.

COWSLIP, PAIGLE *Primula veris* Pl. 13

A downy perennial with obovate leaves, tapering downwards and deeply channelled, down which moisture flows to the roots. The leaves are hairy on the underside. The drooping flowers are borne in umbels of 6–8 on stalks about 6 in. tall with short downy peduncles. They are yellow with an orange-red spot at the base of each petal and are sweetly scented, likened to the breath of a cow or, more so, to a baby's breath. The inflated bell-shaped calyx is of palest green. In bloom April–June.

Habitat: In open meadows and grassy banks, usually where the soil lies over limestone formations as in W Ireland; SE England; Cumberland.

History: One of the most popular of all wild flowers of Europe and the British Isles, beloved of poets throughout the ages. Its sweet scent, present also in the stem and roots, is due to anethol. In the root two glucosides have been separated, one yielding an oil smelling of aniseed, the other smelling of wintergreen. Combined, they produce the scent of the flower.

When in bud and as they open awaiting fertilisation, the flowers release their sweet aniseed scent and become pendulous, providing protection for the nectar and pollen. After fertilisation, the flowers again become erect and remain so until the seed has ripened. Hartley Coleridge observed this characteristic in his lines to 'The Cowslip':

> The coy cowslip, though doom'd to stand
> In state upon the open field,
> Declines her head.

G. B. Andreini has described the flower in these lines:

> How exquisitely sweet
> This rich display of flowers,
> This airy wild of fragrance,
> So lovely to the eye,
> And to the sense so sweet.

It is a plant which denotes good husbandry and which Shakespeare associated with the fragrant Burnet and sweet Clover as in the speech of the Duke of Burgundy in *Henry V*:

> The even mead, that erst brought sweetly forth
> The freckled cowslip, burnet and green clover
> Wanting the scythe, all uncorrected, rank,
> Conceives by idleness . . .

Shakespeare again referred to the Cowslip's freckles in *A Midsummer Night's Dream* when the Fairy chants:

> In their gold coats spots you see,
> Those be rubies, fairy favours,
> In those freckles live their savours . . .

The playwright believed that the Cowslip's scent was given off by the red spots. During Elizabethan times, a delicious wine was made from the flowers, and the leaves were often used in salads. The making of cowslip balls was a favourite pastime of children during May.

In *The Compleat Angler*, Isaak Walton describes the fields of Dovedale when the Cowslip is in bloom: 'Looking down the meadows I could see here a boy gathering lilies and lady's-smocks, and there a girl cropping culverkeys and cowslips, all to make garlands suitable to the present month of May; these, and many other field flowers so perfumed the air, that I thought that very meadow like that field in Sicily of which Diodorus speaks, where the perfumes arising from the place make all dogs that hunt in it to fall off, and to lose their hottest scent'.

FALSE OR COMMON OXLIP *P. veris* × *P. vulgaris*
A natural hybrid which occurs wherever the Primrose and Cowslip appear together. It differs from the true Oxlip, *P. elatior*, in that its umbel is not 1-sided. The first flowers appear singly like Primroses and those appearing later are borne in umbels like those of the Cowslip. The larger individual blooms and loosely-formed umbel also distinguish it from the true Oxlip. It enjoys a more open situation like the Cowslip, rather than shade enjoyed by the true Oxlip. The funnel-like corolla has 5 bosses which are absent in *P. elatior*. Again, the eye is folded, whereas in *P. elatior* it is trumpet-shaped. In bloom April–June.
Habitat: The False Oxlip is to be found in open fields and downland pastures where Cowslips abound. Jack-in-the-Green and Hose-in-Hose forms may occasionally be found (see *P. vulgaris*, p. 139).
History: It was illustrated and described by the French botanist De L'Ecluse (Clusius) in his important *Rariorum Plantarum Historia* (1601), written while he was Professor of Botany at Leyden University. He describes it as *Primula veris pallida flore elatior*, the Large Pale-flowered Cowslip which Gerard also identified when he wrote that 'the flowers are not so thick thrust together (as those of the Cowslip and true Oxlip)'. The flowers are yellower than the buff-coloured ones of *P. elatior*.

The False Oxlip is famous as a parent of the Polyanthus to which it has passed on its sweet perfume. Early in the seventeenth century, the *Turkie-purple Primrose* had been introduced from the Caucasus by John Tradescant. It was known to Parkinson as the *Red-flowered*

Primrose, Primula vulgaris rubra, and it would seem that when crossed with the native hybrid Oxlip, it produced a Red Oxlip or Polyanthus (from the Greek *polyanthos,* many-flowered). It was first mentioned in 1665 by John Rea, a nurseryman of Bewdley in Worcestershire, in his *Flora, Ceres and Pomona.* Rea called it the *Red Cowslip* or *Big Oxlip* 'bearing many flowers on one stalk, in fashion like those of the field but of several red colours; some deeper, others lighter; some bigger like oxlips'. Rea also described the Red Oxlip as being 'more esteemed than those of our own country'. It is interesting that the red colouring proves dominant whenever a Red Polyanthus is crossed with a Polyanthus or Cowslip bearing a yellow or white flower and it was not until the Red Primrose was introduced that the Red Polyanthus could appear. The Cowslip bears a red flower when it is crossed back with the Red Polyanthus. The crossing may have been taken a stage further, with the Turkie-purple Primrose once again, to produce a plant bearing primrose-like flowers of crimson-red, borne in a loose umbel or truss.

In 1683, some twenty years after the publication of Rea's book, the Rev. Samuel Gilbert wrote in his *Florist's Vade Mecum,* 'there are several oxlips or polyanthuses; I have very large hose-in-hose of deeper or lighter reds'. These plants may have been obtained from Rea, for Gilbert was married to Rea's daughter, Miranda, and it was she who had inherited the nursery upon the death of her father in 1681. Shortly after the publication of Gilbert's book, John Evelyn (founder of the Royal Society) referred to the Polyanthus. In 1647, Evelyn had married the twelve-year-old daughter of the English Ambassador in Paris and she was soon to inherit Sayes Court in Kent. In 1687, Evelyn published his *Directions for the Gardener at Sayes Court* in which he mentioned the use of the Red Polyanthus for spring bedding. The following year appeared the first illustration of a Polyanthus, in the *Catalogue* of the Leyden Botanical Garden in Holland and was of a plant obtained from the Botanic Garden at Oxford. Fifty years later the first illustration of a Polyanthus in colour appeared and again the flower was of red colouring.

In 1757 appeared John Hill's famous work *The Vegetable Kingdom.* The twenty-five volumes had taken sixteen years to write, leaving Hill so impoverished that he took to acting as a profession and later became a quack doctor. A coloured illustration shows a Polyanthus of crimson-purple colouring (Turkie-purple) with a bright yellow eye surrounded by a circle of white. Careful inspection also reveals a thin wire-edge of gold from which we may deduct that this was the beginning of the Gold-laced Polyanthus which, for the next hundred years, was to become the most important of the florist's flowers. It survives to this day and has the powerful scent of *P. auricula* which may figure in its parentage.

The now so common Yellow Polyanthus, as we know it today, did not appear until 1880 when Miss Gertrude Jekyll discovered a plant, entirely by chance, growing in her garden at Munstead in Surrey. By the turn of the century, Miss Jekyll's Munstead Strain bearing white and yellow flowers had become firmly established and is obtainable to this day.

The most highly scented of modern Polyanthus strains is the F_2 Hybrid strain raised in Lincolnshire by Mr Harold Hansen. The plants show exceptional vigour, the individual blooms being as large as a ten pence piece and borne in heads like a Cineraria. The flowers are obtainable in shades of pink, apricot, scarlet, blue and gold and when in bloom release so powerful a perfume from an acre field that it may be noticed from a distance of a mile. Like most of the modern strains, the seed is saved from hand-pollinated plants when the 'muddy' crimson and magenta shades of former days are replaced by a brilliance and clarity of colouring.

OXLIP   *P. elatior*   Pl. 13

It has leaves like those of the Cowslip but longer, growing as it would appear from a short stem. The flowers are borne on a long scape, in a 1-sided umbel, but, unlike the Cowslip's, stand erect. The flowers are larger than those of the Cowslip and are buff-yellow with the delicious scent of ripe apricots. The corolla-tube is without the five bosses present in the Cowslip, Primrose and hybrid Oxlip.

The plant loves shade and Shakespeare must have known the true Oxlip when he included it with the Violet – 'Where oxlips and the nodding violet blows', though it is doubtful whether the plant grew about Stratford-on-Avon. In bloom April–May.

Habitat: To be found from earliest times in woodlands and hedgerows in Essex near the village of Great Bardfield, hence its name, Bardfield Oxlip. Occasionally present in the neighbouring county of Cambridge-shire but not elsewhere.

PRIMROSE   *P. vulgaris*   Pl. 13

A perennial with wrinkled lanceolate leaves, downy on the underside. The flowers are borne on short peduncles and appear as if growing singly rather than in loose umbels. They are mostly palest yellow, with a reddish-pink form native to Wales, and measure about 1 in. across. They have the sweet mossy scent of the woodlands. Many forms are to be found, including Jack-in-the-Green, Hose-in-Hose, etc. The calyx has 5 bosses or plaits and acute teeth. The corolla-tube is contracted at the mouth, while the petals are attractively notched. In bloom March–June.

Habitat: Prolific in woodlands and hedgerows throughout the British Isles, especially in the west of England and Ireland and in the Midland counties. Also found on railway embankments where they cannot be easily disturbed by plant hunters.

History: The most loved of all wild flowers because of its earliness to bloom and its hardiness. Robert Burns wrote: 'The primrose I will pu', the firstling o' the year', and John Clare in *The Shepherd's Calendar* for March, written at Helpston in 1822, conveys his joy at the approach of spring and the appearance of the first primrose:

> A tale of spring around the distant haze
> Seems muttering pleasures with the lengthening days;
> Morn wakens, mottled oft with May-day-stains;
> And shower-drops hang the grassy sprouting plains,
> Or on the naked thorns of brassy hue
> Drip glistening, like a summer-dream of dew.
> The woodman, in his pathway down the wood,
> Crushes with hasty feet full many a bud
> Of early primrose; yet if timely spied,
> Shelter'd some half-rotten stump beside,
> The sight will cheer his solitary hour,
> And urge his feet to stride and save the flower.

To the earlier poets it was the first in esteem of all flowers. In Spenser's *Daphnaida*, the husband lamenting the loss of his young wife says:

> Mine was the Primrose in the lowly shade! . . .
> Oh! that so fair a flower so soon should fade . . .

Chaucer, in *The Miller's Tale*, described the 'fair lady' as a *Prime-role* and this was the way the name was spelt until the end of the sixteenth century when Shakespeare so often mentioned the flower, usually referring to its pale, fragile quality, as in *Cymbeline*, perhaps with the face of his dead son Hamnet in mind:

> Thou shalt not lack
> The flower that's like thy face, pale Primrose.

Again, in *A Midsummer Night's Dream*:

> In the wood where often you and I
> Upon faint primrose beds were wont to lie.

Keats in his *Endymion* conveys similar feelings:

> . . . Their smiles
> Wan as primroses, gather'd at midnight
> By chilly finger'd spring.

The Primrose's pale colouring is also mentioned in Milton's poetry – the only quality of the flower he referred to:

. . . The flowery May, who from her green lap throws
The yellow cowslip and the pale primrose.

Coleridge, like Clare, expressed his happiness upon finding the first
flowers:

In dewy glades,
The peering primrose, like sudden gladness,
Gleams on the soul.

'This Sweet Infanta of the yeare' was how Robert Herrick described the
flower in his poem 'To the Primrose'.

In earliest times, the Primrose was cultivated entirely for its medicinal
and sweetening qualities. Parkinson mentions that the juice of the stems
rubbed on the face removed unsightly spots and 'next to Betony was
used to relieve pains in the head'. Thomas Tusser listed the Primrose
with 'herbs for the kitchen' and Lyte said that 'primroses are now used
dayly amongst other pot herbs'.

As late as the early years of the nineteenth century the Primrose was
still cultivated for its medicinal qualities. In *The Family Herbal* written
by Sir John Hill, M.D., and published in 1812, we read, '. . . The juice of
it (the root) snuffed up the nose occasions sneezing and (as Parkinson
said two centuries before) is a good remedy against headaches. The
dried root powdered has the same effect, but not so powerfully'.

The candying of Primroses and Violets was popular during the
seventeenth century and in her *Accomplished Lady's Delight* (1719),
Mary Eales, confectioner to Queen Anne, described how this was done
by steeping gum arabic in water and wetting the fresh flowers with the
solution. They were then dipped in castor sugar and hung up in a warm
room to dry.

The first writer to mention the Double Primrose was Tabernae-
montanus who wrote of the Double Yellow variety in 1500. It was the
double form of our single native Primrose, known later as the *Double
Sulphur*, which may have appeared with the Double White at about the
same time. Indeed, natural 'sports' in yellow and white are quite
common, and in *Country Life* for 27th June, 1957, a photograph
appeared of the Double White growing in a wood in Kent.

Gerard mentioned in particular this same Double White form, *alba
plena*, in his *Herbal* (1597) and wrote, 'our garden Double Primrose, of
all the rest is of the greatest beauty'. In his *Catalogue of Plants* he lists
both the Single Green and the Double Green Primrose, and writing
thirty years later, Parkinson gives a full description of Double Primroses
which were then popular garden plants. He describes the flowers as
being 'very thick and double and of the same sweet scent with them (as
the common single field Primroses)', though very few of the Double
Primroses have any degree of perfume and the old Double White,

Double Yellow and Quaker's Bonnet, which is believed to be a 'sport' of *P. rubra* and is among the loveliest plants of the garden, are entirely devoid of any scent. Being natural 'sports' they are, however, among the most vigorous of the doubles, persisting to this day.

Some heavily scented Double Primroses:

BON ACCORD CERISE. One of a dozen Double Primroses raised in Aberdeen early this century which take their name from the motto of the city of Aberdeen. They are what is known as 'bunch' Primroses, with some flowers borne on footstalks like the native Primrose and others, from the same plant, on short polyanthus stems. All are in cultivation and several bear scented flowers; Bon Accord Cerise has the powerful sweet honeysuckle perfume of the Auriculas.

The blooms are over an inch wide and are rosette-shaped with rounded petals and have the appearance of a small Carnation, with colouring of a clear cerise-pink.

CRATHES CRIMSON. This was found in the grounds of Crathes Castle in Aberdeenshire but appears to be the long-lost Bon Accord Brightness, as it is the same shape and has the yellow shading at the base of the petals common to all Bon Accords. The neat, rounded button-like flowers are of pale crimson-red with a powerful scent.

MARIE CROUSSE. This was introduced from France about the year 1800 and is one of the most satisfying spring-flowering plants. It is a strong grower and blooms in profusion, increasing with each year. The foliage is dull green, and it bears fully double blooms on short polyanthus stems, an inch or so wide and of a lovely shade of Parma violet, splashed and edged with white. They have a powerful honeysuckle perfume. It received the Award of Merit from the Royal Horticultural Society in 1882.

RED PADDY. This is the *P. rubra plena* of Irish cottage gardens, possibly a 'sport' of the old Turkie-purple, *P. rubra*, the so-called 'red' Primrose of Eastern Europe. It bears the same neat, flat and symmetrical flower as Crathes Crimson and is of similar colouring. It would seem that they have a common ancestor for they also have the same sweet perfume. In this case the rosy-red blooms have an attractive wire edge of silver.

There are a number of lovely forms of the wild Primrose including Jack-in-the-Green. Here the flowers are backed by a Tudor ruff of slightly larger green leaves, which persists long after the flowers are dead. One, named Eldorado, has the Primrose perfume, the blooms being large and a brilliant golden-yellow. It is almost certainly a derivative of *P. vulgaris*.

Another attractive form is the Hose-in-Hose which Gerard called Two-in-Hose, and of which Parkinson wrote, 'they remind one of the breeches men do wear'. They were also known as *Duplex* or *Cup-and-Saucer* Primroses, for one bloom can be seen growing out of another like the hose worn by gentlemen of the Elizabethan period, with one stocking pulled up to the thighs and the other turned down immediately below the knees, to give the appearance of one growing out from another. The botanical explanation of this delightful fairy-like effect is that the lower bloom is really a petaloid calyx. This is the Primrose described by Mrs Ewing in *Mary's Meadow*.

Almost all the Hose-in-Hose are derivatives of the Common Primrose and diffuse a sweet mossy perfume which is, however, more pronounced than the scent of *P. vulgaris* for there are two blooms instead of one and both are scented. The Hose-in-Hose form of *P. juliae* Wanda is entirely without perfume as would be expected, for *P. juliae* of the Caucasus is scentless.

Some varieties of Hose-in-Hose

BRIMSTONE. Of semi-polyanthus form, the large bell-shaped blooms are borne on 4-in. stems and are of a lovely clear sulphur-yellow colouring.

CANARY BIRD. The flowers, borne one inside the other, are a bright canary yellow.

GOLD-LACED HOSE. A Hose-in-Hose form of polyanthus habit, the dark crimson-black flowers held on 6–8-in. stems, each bloom being clearly edged with gold. The fragrance is even more pronounced than in the single gold-laced form.

IRISH MOLLY. Also known as Lady Molly, it is a hose of true Primrose form and bears large mauve-pink blooms which are delicately scented.

LADY DORA. A delightful old Irish form bearing small, dainty blooms of brilliant golden-yellow with the scent of honeysuckle.

LADY LETTICE. It comes into bloom early, covering itself in a fairy-like mass of dancing apricot-coloured blooms, flushed with salmon and pink and with a delicious sweet perfume.

Another interesting form of Primrose is the Jackanapes, in which the leaves of the ruff are striped with the colouring of the bloom – or the ruff may be said to be of the same flower colour, striped with green. It takes its name from a striped coat fashionable during the seventeenth century. In his *Diary* for 5th July, 1660, Samuel Pepys wrote, 'This morning my brother Tom brought me my Jackanapes coat'.

The Jackanapes-on-Horseback is yet another quaint form. It is the *Franticke* or *Foolish Primrose* so well described by Parkinson in the

Paradisus. 'It is called Foolish', he wrote, 'because it beareth at the top of the stalk a tuft of small green leaves with some yellow leaves, as it were pieces of flowers broken and standing amongst the green leaves'. It is illustrated in the *Paradisus,* the leaves being larger, coarser and more widely arranged than those of the Jack-in-the-Green form. A book of early Stuart times describes them as 'all green and jagged', while Gerard said that it was so named by the women, 'the flowers being wrinkled and curled in a most strange manner', though it is not clear what it has to do with 'horseback'.

Another strange form is the Gally Gaskin. This is an ordinary Single Primrose with an especially swollen calyx and with a frilled ruff beneath.

CYCLAMEN Corm-bearing perennials with radical, petiolate leaves and nodding flowers, borne solitary on erect leafless scapes, twisting spirally inwards when in fruit and in certain species depositing the fruit directly into the soil. Flowers usually purple or pink; corolla-tube long, petals reflexed. Stamens inserted at base of tube. Several species scented and pollinated by Lepidoptera.

COMMON SOWBREAD *Cyclamen hederifolium*
Syn: *C. europaeum, C. odoratum*
It forms a large flat globular corm which will increase in size for 20 years or more, until it measures 12 in. across. At the top of the corm appears a tuft of heart-shaped leaves of darkest green, mottled with purple and white. The leaves appear at the same time as the flowers which hover above the corm like pink butterflies held on 3-in. stems. They have a soft, primrose-like perfume though the rare white form is more heavily scented. When planted, the upper surface of the corm should be exposed. In bloom August–September.
Habitat: Preferring shade, it is to be found today in woodlands in SW England and Wales and in Kent, always growing in a limestone soil. Now very rare.
History: If not a true native, it may have reached Britain with the Romans for it abounds in S Europe, and the corms were used as fodder for pigs. It was also much in demand for easing childbirth. Gerard tells that the plant 'grew upon the mountains of Wales and on the hills of Lincolnshire and Somerset'. It was also found naturalised in Kent and Sussex.

BUDDLEIA FAMILY Buddleiaceae

Trees or shrubs with mainly opposite simple leaves, stipulate or ex-stipulate and covered in glandular hairs. Flowers hermaphrodite, 4-merous; stamens alternating with corolla-lobes. Ovary of 2 united carpels. Fruit, a capsule or berry.

Plate 13
Primrose Family 1 Cowslip 2 Oxlip
3 Primrose

Plate 14
Broomrape Family 1 Clove-scented Broomrape (a parasite on Hedge Bedstraw, 2, which is also shown)

BUDDLEIA Trees or shrubs covered in glandular hairs and with pithy stems. Leaves lanceolate, woolly on the underside. Flowers borne in dense panicles or cymes. Calyx campanulate; corolla with straight cylindrical tube with the stamens inserted, pubescent on the outside. Fruit, a capsule.

BUTTERFLY-FLOWER *Buddleia davidii*

A deciduous shrub with pubescent, pithy wood, bright brown in colour and somewhat brittle. The dark green leaves are lanceolate and serrated and covered in silky white hairs on the underside. The flowers are lilac-pink, borne in dense narrow panicles and have a musky 'summery' smell, resembling warmed honey. Pollination is almost entirely by butterflies which descend on to the long graceful panicles on sunny days, extracting the honey from the long tubes. In bloom July–August.
Habitat: Often found naturalised on old walls, railway embankments and waste ground in or near towns as a garden escape.
History: Mostly native to tropical SE Asia and closely related to the strychnine family of plants, the Buddleia was named in honour of Rev. Adam Buddle and *B. davidii*, after Abbot David, a missionary in S China who first obtained seed. It is a shrub of robust constitution which grows best in a sandy soil close to the sea. A number of magnificent garden hybrids bearing elegant flower-spikes in the richest of colours have brought the plant much popularity in the modern garden. Among the finest are Black Knight, bearing panicles of deepest violet-purple with a pronounced sweet scent; Royal Red, crimson-purple; White Bouquet; and Glasnevin Blue, whose beauty is enhanced by its grey foliage. All are honey-scented.

OLIVE FAMILY Oleaceae

Erect or climbing trees or shrubs. Leaves opposite, exstipulate, simple or pinnate. Flowers borne in cymose inflorescence or panicle. Calyx 4-cleft, inferior, persistent or absent; corolla of 4 united or free petals; stamens 2; ovary 2-chambered. Seeds with or without endosperm. Mostly native of S Europe and Asia, the flowers are usually white or pale yellow with a sickly sweet scent, e.g. *Syringa, Olea, Osmanthus*. Pollination by Lepidoptera and Hymenoptera, for the nectar is concealed at the base of a long tube. One only, *Ligustrum*, is native of the British Isles and has scented attractions.

LIGUSTRUM Trees or shrubs with opposite, entire leaves, evergreen or deciduous. Flowers hermaphrodite, borne in 3-forked terminal panicles or in clusters. Calyx 4-toothed; corolla funnel-shaped; 4-lobed. Stamens 2. Fruit, a drupe with 1–2 seeds.

COMMON PRIVET *Ligustrum vulgare*

An almost evergreen shrub growing up to 10 ft tall with glossy, dark green lance-shaped leaves. Flowers white, borne in panicles, the calyx being cup-shaped, the corolla funnel-like, 10 mm long with 4 spreading lobes. The honey, secreted at the base of the tube, is sheltered from adverse weather by 2 stamens which almost fill the mouth of the tube and protect the honey from short-tongued insects. The scent is sweet but, due to the presence of trimethylamine, has an ammonia undertone less concentrated but similar to that in *Sorbus, Cotoneaster*, etc., of the Rosaceae family. The fishy undertone is absent from the closely related Lilac and Jasmine. Honey from bees which have fed on Privet blossoms has a fishy taste and smell. The flowers are followed by shining black fruits in autumn. In bloom June–July.

Habitat: In hedgerows and woodlands throughout England and Wales, usually in a limestone soil.

History: It takes its name Privet from the word *prim*, as it was a plant well suited for a cottage hedge and could be kept 'prim' and tidy. Also it is almost evergreen. Parkinson tells us that it was used to cover the sides of arbours and bowers and adds, 'it is cut into many forms of men, horses, birds, etc.' A distilled water was obtained from the berries and leaves, which was used as a cure for ulcers of the mouth and throat, while the pulp of the berries produced a crimson dye which, with the addition of alum, was used for dying silk a durable deep green.

BORAGE FAMILY   Boraginaceae

Annual or perennial herbs, trees or shrubs with mostly alternate, exstipulate leaves covered in stout hairs. Flowers, which are usually hermaphrodite, unroll to reveal a 1-sided raceme or panicle. Calyx bell-shaped or absent, with 5 teeth or lobes. Corolla funnel- or salver-shaped with 5 (rarely 4) or 6–8 lobes; stamens equal to corolla-lobes and alternating. Ovary usually 2-, sometimes 4-celled. Fruit 2–4 nutlets; seed without endosperm.

Almost all bear brilliant blue flowers, often with a yellow eye or direction lines, and their blooms are borne pendulous with the abundant honey deeply concealed. They are almost all pollinated by Hymenoptera but self-fertilisation takes place in the absence of bees. Few of the species have scented flowers, with the exception of the Alpine Forget-me-not, and scent is necessary in the yellow-flowered *Onosma* to attract pollinating insects which are scarce at the higher altitudes in which these plants grow. The juice of the leaves and stem of Borage smells and tastes of cucumber.

CYNOGLOSSUM Annual or perennial plants covered in silky

hairs. Leaves lanceolate; flowers borne in forked cymes. Calyx 5-cleft; corolla funnel-shaped with scales in the mouth. Nutlets 4, covered with hooked bristles.

HOUNDSTONGUE
Cynoglossum officinale

A perennial (or biennial) plant growing up to 2 ft tall with broad, lanceolate leaves, grey in appearance due to the covering of downy hairs, which in shape and appearance resemble a hound's tongue. Due to the presence of esters of fatty acids, they emit a fur-like smell, similar to that of the Lizard Orchid, when handled. The flowers are large and purplish-red and are borne in branched cymes. They have crimson-red scales in the mouth. The flattened nutlets are covered in bristles or spines which adhere to the clothes or to the coats of animals upon which they rely for their distribution. In bloom June–August.
Habitat: Widespread on waste ground and on cliffs of chalk or limestone.

Houndstongue

BORAGO Annual or perennial herbs covered in bristles. Leaves ovate. Flowers borne in forked cymes; calyx 5-cleft; corolla rotate with 5 scales in throat. Stamens in corolla-throat; filaments broad and forked; anthers converging in a cone. The pendulous flowers secrete honey at base of ovary. Pollination by bees.

BORAGE *Borago officinalis*

An annual plant growing 2 ft tall, with all parts covered in bulbous hairs. The leaves are oval and greyish-green; the flowers, brilliant blue with purple-black anthers, are borne in forked cymes. The corolla-lobes alternate with the lanceolate calyx-teeth. Nectar is secreted at the base of the pale yellow ovary. The honey is held in a short tube formed by the base of the stamens. The anthers meet cone-like and each dehisces slowly, allowing the pollen to fall on to the cone within which the style and stigmas are enclosed. When an insect thrusts its proboscis between the stamens, the anthers are displaced and the top of the cone opens to allow the pollen grains to fall on to the stigmas, thus effecting

BORAGE FAMILY

fertilisation. The flowers are unscented but when the stem or leaves are broken the juice of the plant releases a pleasant moss-like or cucumber smell. In bloom June–August.

Habitat: Possibly a native plant or an early introduction; usually found on waste ground where the soil is of a sandy nature.

History: Besides its pleasant smell when released by pressure, the juice has a cooling effect in drinks and is today used for this purpose, as it was in earlier times to flavour cider and claret cup. The leaves were also used in salads for they were considered to be a valuable tonic. In his *Herbal Simples* (1895) Dr Fernie of Glasgow said that its reputed powers of invigoration could be medically substantiated for the juice contained 30% nitrate of potash. Countrymen used to place the leaves with cheese in sandwiches to provide lasting nourishment. The name is a corruption of *cor-ago*, *cor*, the heart and *ago*, I stimulate. There is an old Latin adage which, when translated, runs, 'I, Borage, bring always courage' and Evelyn wrote, 'sprigs of borage are of known virtue to revive the hypochondriac and cheer the hard student'.

It was named *Euphrosynon* by the Greeks, because when put into wine it gave the drinkers a feeling of well-being. A valuable winter cordial is made by simmering the leaves, while the flowers may be candied, like Violets, and used in confectionery.

In the household book of the Earl of Northumberland for the year 1502, there is a list of herbs 'to stylle' for making sweet waters which were given as presents on Saints' days and birthdays. Those plants used in the still room, which was to be found in most households of importance, included 'borage, columbine, bugloss, sorrel, cowslip, scabious (*Knautia arvensis*), tansy, wormwood, sage, dandelion and hart's-tongue'.

MYOSOTIS Hairy annual or perennial herbs with sessile cauline leaves. Flowers mostly blue, borne in a 1-sided raceme; calyx 5-cleft; corolla salver-shaped, lobes blunt with scales in throat. Nutlets smooth. Flowers of most species scentless and pollinated by Hymenoptera and also by Diptera, for the flowers are conspicuous by aggregation. Self-fertilisation is also possible in absence of insects. Takes its botanical name from the resemblance of the leaves to the ears of a mouse.

Alpine Forget-me-not

ALPINE FORGET-ME-NOT *Myosotis alpestris*

A pubescent perennial growing from a rhizomatous rootstock, the stems covered in stiff hairs. The oblong-lanceolate lower leaves have long petioles, the upper being sessile and also hairy. The deep purple-blue flowers are borne in compact clusters on 3-in. stems. The calyx is campanulate with narrow teeth, the tube being about ¾ in. long, the corolla-lobes flat and round. In the high alpine regions, the flowers need scent to attract the few insects about, and the flowers are deliciously fragrant. It is pollinated almost entirely by Lepidoptera during the day. In bloom July–August.

Habitat: Very rare but present in several mountainous regions of N Yorkshire; on Mickle Fell in Westmorland; and on Ben Lawers in Scotland, usually at a height of between 3000 and 4000 ft.

BINDWEED FAMILY Convolvulaceae

Herbs or shrubs, often climbing, which inhabit the tropical and temperate regions of the world. Leaves alternate, exstipulate. Milky juice (latex) usually present in stems and foliage. Flowers large, trumpet-shaped, with abundance of honey secreted at base of tube. Pollinators include humming-birds (tropics); Hymenoptera; Diptera. Flowers mostly white or blue, very few scented. Family includes parastic climber *Cuscuta*.

CONVOLVULUS Annual or perennial herbs, often climbing and with stems woody at base. Leaves alternate, exstipulate. Flowers trumpet-shaped; sepals 5, inferior, imbricate; corolla 5-toothed; stamens, 5 inserted at base of corolla tube. Ovary, 2–4 united carpels; style single, 2–4 forked. Fruit a 1-chambered capsule.

FIELD OR LESSER BINDWEED
Convolvulus arvensis

A straggling or climbing perennial growing from a stout rhizomatous root and with stems which climb 2 ft

Field Bindweed

high in an anticlockwise direction. The leaves are arrow-shaped and pubescent when young, with a short petiole. The flowers are borne 1–3 together and are white, sometimes striped with green, and occasionally rosy-red with white radiating lines. The corolla is yellow at the base on the inside where the honey is secreted. This is protected by the base of the stamens and is obtained through 5 narrow apertures. The flowers emit a soft almond-like scent on a sunny day, which is intensified by the flowers' closing tightly in dull weather and at night. They are visited by all types of insect, for even Diptera are able to climb down the wide tube and reach the honey. In bloom June–October.

Habitat: In cornfields throughout the British Isles; also in sand quarries and by roadsides, especially in the Midlands and where the soil is sandy and well drained.

History: Distinguished from the Greater Bindweed, *Calystegia sepium*, which is scentless, by the bracteoles which envelop the calyx. It is Wordsworth's 'Cumbrous bindweed, with its wreaths and bells'. Gerard confused it with *Calystegia sepium* by saying that the Greater Bindweed was 'sweet of smell'. The genus takes its name from the Latin *convolvo*, I entwine, but it is not nearly so vigorous as the Greater Bindweed, as it trails rather than climbs.

NIGHTSHADE FAMILY   Solanaceae

Herbs, shrubs or small trees, many of which are highly poisonous narcotics. Most are native to the tropics of South America, and among them are the Potato and Tobacco (*Nicotiana*) plants, the latter bearing night-scented flowers and pollinated by Lepidoptera. Several are evil-smelling plants. Leaves alternate or in pairs. Flowers borne solitary or in cymous inflorescence; calyx inferior, deeply cleft; corolla 5-lobed; stamens alternating with corolla-lobes. Ovary 2-chambered; style 1; stigma simple. Fruit a 2–4-chambered capsule or berry. Closely related to the almost scentless Scrophulariaceae family. Flowers often a lurid yellow-brown with purple-brown anthers, pollinated by Diptera; occasionally by humble-bees.

HYOSCYAMUS Annual or biennial herbs, branched at base. Leaves alternate, often pinnatifid, exstipulate. Flowers borne in loose axillary cymes; calyx 5-toothed; corolla funnel-shaped, with unequal lobes. Stamens inserted at base of tube. Ovary 2-celled; stigma capitate. Fruit a capsule, opening transversely.

HENBANE, HOGBEAM   *Hyoscyamus niger*
A sticky pubescent annual or biennial growing up to 4 ft tall with large woolly leaves. Those near the base have teeth. The flowers are funnel-

shaped, cream-coloured with purple veins and arranged in a double row. They are followed by a 2-chambered capsule. The whole plant gives off an unpleasant sickly, fetid smell of dead rats, possibly an endowment of nature to deter humans from approaching too near, for all parts are highly poisonous. In bloom June–August.

Habitat: On waste ground, usually of a chalky nature and often near the sea; also on railway banks and farmsteads throughout the British Isles.

History: 'The whole plant', wrote Culpeper, 'hath a very ill, odoriferous smell'. It may have provided the poisonous juice, the 'cursed hebenon', a volatile alkaloid, which killed Hamlet's father. Gerard said that merely to smell the flowers brought about heavy sleep and to take the seeds internally 'causeth an unquiet sleep, which continueth long'.

FIGWORT FAMILY Scrophulariaceae

Mostly herbs, few trees or shrubs. Leaves alternate, opposite, whorled, exstipulate. Several climbing, several parasitic. Flowers borne in cymose inflorescence, axillary or terminal. Calyx 4–5-lobed; corolla 2-lipped; stamens 4 or 5; alternating with corolla-lobes. Ovary 2-chambered; style 1; stigma 2-lobed. Fruit a 2-chambered capsule. Seeds with endosperm. Many, like Solanaceae, contain poisonous juices, the entire family being almost scentless. The flowers are predominantly yellow and are pollinated mostly by Hymenoptera, especially humble-bees which alight on them and because of their weight are able to open the lip, as in *Linaria, Antirrhinum.*

VERBASCUM Erect perennial or biennial herbs with scattered leaves covered in down. Flowers borne in terminal racemes; calyx 5-lobed, rotate; stamens 5, hairy. Fruit a 2-valved, many-seeded capsule. Minute quantities of honey are secreted and pollination is by Hymenoptera and occasionally Lepidoptera, though the flowers are scentless. Self-fertilisation also takes place in absence of insects.

COMMON MULLEIN, AARON'S ROD *Verbascum thapsus*

A stout herbaceous biennial reaching a height of 4–5 ft, the stem and lanceolate leaves begin covered in white downy hairs. The yellow flowers are borne in an extended raceme; the bracts are longer than the flowers; the sepals triangular. Two stamens are longer than the others with their filaments covered in white hairs. The anthers are orange-coloured, the stigma capitate. The leaves have a refreshing fruity scent when handled, resembling *Rosa rubiginosa.* In bloom June–September.

Habitat: By the wayside and on sunny banks of sandy soil throughout the British Isles.

History: At one time its leaves were in demand for the relief of pulmonary diseases and were at their best gathered just before the appearance of the flower spike. Moreover, dried and smoked in a pipe they will cure a stubborn cough. The plant takes its country name *Figwort* (given to each genus in the family) from the use of the leaves to wrap around figs, which kept the fruits in a moist condition for several months. Another old name was *Hedge Taper* or *Torch-plant*, for the woolly stems (and leaves) were dipped in tallow or suet and burnt to give light at outdoor country gatherings or in the home. For this reason the plant was named *Candelaria* by the Romans. Its botanical name is a derivative of the Latin *barbascum*, a beard, an allusion to its woolly hairs. Until recent times, the fresh flowers were steeped in olive oil for 3–4 weeks and exposed to the sunlight, and 2–3 drops of this oil in the ear would cure the most severe pain. The fresh leaves were placed in country children's shoes to cushion their feet against rough roads.

ANTIRRHINUM   Annual or perennial herbs differing from *Linaria* in the pouch placed at the base of the corolla. Lower leaves opposite, upper alternate. Flowers borne in terminal racemes; calyx 5-lobed; corolla 2-lipped with mouth which opens and closes to protect the

Common Mullein   *Snapdragon*

honey. Fruit, a capsule opening by pores. Pollination by bees which alone are able to open the lower lip with their weight. Self-pollination also takes place when the mask falls away to give the seed vessel access to the air.

SNAPDRAGON *Antirrhinum majus*

A perennial herb growing 12–20 in. tall, with woody stalks where long-established. The lanceolate leaves are slightly pubescent. The flowers are borne in a terminal spike-like raceme and have a pouched lower lip which opens when pressed at the sides; the sepals are blunt and shorter than the corolla; the capsule longer than the calyx. In bloom July–October.

Habitat: On old walls, growing in the mortar, and on limestone cliffs and quarries.

History: Perhaps a native plant or possibly introduced by the Normans, for it is found, like the Pink, about ruined castles and on monastery walls as at Ely, and at St Augustine's Abbey in Canterbury. It was well established by early Tudor times for Lyte included it in his *New Herbal* and described its appearance as 'not much unlike the flowers of toadflax (*Linaria*)'. Gerard tells us that he grew the white and purple varieties and added that 'the herb being handed about one, preserveth a man from being bewitched'. Henry Phillips in the *Flora Historica* says that it is 'found on the cliffs of Dover and is classed as one of the native plants of England'. He says that 'on pressing the sides of the flower, it opens like a gaping mouth, the stigma appearing to represent a tongue and upon removing the pressure, the lips of the corolla snap together, hence the children have named it Snapdragon'.

The flowers have a musky scent when warmed by the sun, a smell associated with late summer flowers.

Matthew Arnold, in his *Thyrsis*, included it among the plants associated with midsummer, each of which has a heavy Eastern perfume:

> Soon will the high Midsummer pomps come on,
> Soon will the musk Carnations break and swell,
> Soon will we have gold-dusted Snapdragon,
> Sweet William with his homely cottage smell,
> And stocks in fragrant blow.

In 1964 a trio of sweetly scented F1 hybrid varieties of outstanding beauty appeared, the tall graceful spikes bearing 40 or more large, fully double flowers with attractively ruffled petals. Growing to a height of 3 ft, the plants bear 10–12 spikes in bloom at one time. Super Jet bears apricot-orange flowers; Venus, peach-pink; and Vanguard, deep cerise with a pronounced clove perfume.

BROOMRAPE FAMILY Orobanchaceae

Tuberous-rooted parasitic herbs with chlorophyll absent or almost so, which attach themselves by suckers to the roots of other plants, e.g. Thyme, Broom, Bedstraw. Leaves absent, replaced by pointed scales. Flowers borne in spike or raceme with bract at base of each flower. Calyx tubular, 2–5-toothed; corolla 2-lipped. Stamens 4, arranged in twos. Ovary 1-chambered; style 1; stigma 2-lobed; capsule 2-valved, seeds small. Seeds cannot germinate unless in contact with roots of host plant. Several visited by bees, but self-pollination possible.

OROBANCHE Annual or perennial parasitic herbs bearing underground tubers which attach themselves to the roots of the host plant. The flowers are borne in a 1-sided terminal spike which is covered in scales. Calyx 2-lipped, toothed. Corolla 4–5-cleft, upper lip erect, 2-lobed. Stamens in corolla-tube; style curving downwards at end. A scale or bract is present beneath each flower.

THYME BROOMRAPE *Orobanche alba*

It grows only 6 in. tall and has red stems and flowers borne in a loose raceme which have a pleasant sweet scent. It is named *alba* because in Europe its flowers are almost devoid of colour. The stem is furnished with red scales and is covered in glandular hairs. The 2-lipped calyx equals the corolla-tube, the upper lip being notched and spreading. The stamens with their hairy filaments are contained in the corolla-tube. In bloom June–July.

Habitat: Parasitic on Wild Thyme and present on rocky hillsides and downland, usually close to the sea and especially in Cornwall and NE Yorkshire.

CLOVE-SCENTED BROOMRAPE *O. caryophyllacea*
Pl. 14

An annual with purple-brown stems some 8 in. tall, with scales of similar colouring. The flowers are pink and are borne in a loose spike. They have a distinct clove-like scent which is most pronounced during warm weather. The flowers are densely pubescent. Nectar is present at the base of the corolla-tube and the flowers are visited by bees and butterflies. In bloom June–July.

Habitat: Attaches itself to Hedge Bedstraw, *Galium mollugo*, and is found only in SE Kent, close to the sea and in sandy soil.

THYME FAMILY Labiatae

Cosmopolitan throughout temperate regions but mostly Mediterranean with several genera native to or long naturalised in British Isles.

Perennial herbs or undershrubs with square stems and opposite, ex-stipulate leaves. Flowers 2-lipped, borne in axillary cymes, or solitary or in pairs. Calyx 5-toothed or 2-lipped; corolla 2-lipped and usually 5-lobed with 2 upper lobes united into a lip and the other 3 united to form a lower lip. Stamens 4, alternating with corolla-lobes. Ovary superior, 4-lobed or parted; style 2-branched. Fruit 4, 1-seeded nutlets enclosed in persistent calyx.

Throughout the family, the flowers are blue, purple-blue or pinkish-mauve and are scentless. Only the leafy parts of the plant possess the familiar aromatic herby scent and this is due to an essential oil which is present in glands in the leaves and bracts and which is released upon bruising or by drying. Pollination almost entirely by Hymenoptera; to a lesser extent by Lepidoptera. Of those genera not native to the British Isles, Patchouli and Lavender are grown commercially for their scented attractions, the latter in Surrey.

In Lavender, the essential oil is composed of the bergamot-scented ester, linalyl acetate and eucalyptol, the latter combining with citral to produce the strong lemony scent of Lemon Thyme. In Lavender, linalol is present in the immature bud and combines with acetic acid in the open bud to produce the pleasantly scented linalyl acetate. The essential oil is present in all green parts of the plant, including the leafy calyx, but not in the petals. Borneol acetate, smelling of pine (and present in pine needles) is found in Rosemary, while menthol is present in Mint, camphor and eucalyptol in Sage and Thyme. Eucalyptol, present in varying amounts in all the Labiates, is a powerful antiseptic. The essential oil of Thyme has an antiseptic value twelve times more powerful than that of carbolic acid, while that of Rosemary is five times as powerful, hence the esteem in which these plants were held in earlier times.

MENTHA   Perennial herbs with a creeping rhizomatous rootstock. Leaves lanceolate, sometimes toothed. Flowers purple, pink or white, borne in axillary or terminal whorls. Bracts numerous. Calyx tubular with 5 equal teeth; corolla-tube shorter than calyx. Stamens 4, equal. Nutlets ovoid, smooth.

CORSICAN MINT   *Mentha requienii*
An early introduction which is naturalised in parts of Britain. An almost prostrate perennial with creeping stems, rooting at the nodes. It has smooth, pale-green peppermint-scented leaves and bears tiny pale lilac flowers, hairy in the throat. It grows rampant between crazy paving stones and with the creeping Thymes makes a fragrant 'lawn' which, when trodden upon, releases a most refreshing scent. In bloom June–September.

Habitat: Naturalised on mountainous slopes in N Ireland and Scotland where it enjoys the cool moist climate.

PENNYROYAL   *M. pulegium*   Pl. 15

A prostrate pubescent perennial with oval toothed leaves covered in glandular dots which release a peppermint scent when pressed. The flowers, with their hairy calyx, are pale reddish-pink and are borne in tiny whorls on 3-in. stems. The corolla is hairy only on the outside with the stamens protruding. In bloom August–September.

Habitat: On moist heathlands and pastures and on the banks of brooks, usually growing in sandy soil and mostly confined to S England; SW Scotland; E Ireland; and the Channel Isles.

History: Widely distributed in S Europe, it was named by Pliny, for it was in common use at the time to rid houses of fleas. In Aelfric's *Vocabulary* there is a reference to *Pollegia* which may be taken to mean this plant. Later the name became *Puliall* to which was added *royal* because of its use by royalty in ridding their residences of fleas. Gerard mentions that it was to be found at 'Mile End, near London' where it was sold in the streets, mostly to sailors to sweeten drinking water whilst at sea, and also for ridding their quarters of fleas. It makes a most refreshing bath 'to comfort the nerves and sinews'. It was also worn as a chaplet to prevent giddiness and was used in posies to keep away flies. To Pliny it was the most valuable of all plants and he considered the leaves a help for headaches. Matthiolus suggested using a distillation of the leaves to freshen the eyes and help the sight.

It was also called *Pudding Grass* and was used in stuffings which were known as *puddings*.

CORN MINT   *M. arvensis*

A downy perennial growing 12–18 in. tall with weakly stems and large oval serrate leaves which release the smell of mint, with rancid undertones when handled. The small lilac flowers are borne in whorls, the calyx being bell-shaped and hairy with the stamens protruding. In bloom July–September.

Habitat: In cornfields and hedgerows and at the edges of woodlands throughout the British Isles, especially S England and the Midlands.

History: This is the Mint with which Shakespeare must have been familiar, for Peppermint, though a native plant, was not discovered until about 1700. The playwright mentioned the Corn Mint, with other 'hot' plants in *A Winter's Tale* when Perdita says:

> Here's flowers for you
> Hot Lavender, Mints, Savory, Marjoram:

The Marigold . . . these are flowers
Of middle summer, and I think they are given
To men of middle age.

Culpeper suggests the use of hot rose-petals and mint leaves (applied to the head) as a cure for sleeplessness, while mint leaves alone make a refreshing and relaxing bath. Sprigs of Corn Mint placed among cheeses will 'preserve them from corruption', while bees will never desert a hive that has first been rubbed with Corn Mint.

'Mint I grow in abundance and in all its varieties. How many they are. I might as well try to count the sparks from Vulcan's furnace'. This introduction to the Mints is taken from a translation of a ninth-century poem of the monk Walafrid Strabo who, in *The Little Garden*, described the herbs and flowers he grew during the reign of Charlemagne, in a typical monastic garden on the shores of Lake Constance.

In France, Mint is given the highest honour. It is known as *Menthe de Notre Dame*, and in Italy *Erba Santa Maria*, possibly because in medieval times the herb was so much used for strewing in churches. In the *Polyolbion* Michael Drayton mentioned Balm and Mint as herbs liberally used for this purpose (p. 18).

Pliny said that 'the smell of mint doth stir up the mind and the taste to a greedy desire for meat' and 'it will not suffer milk in the stomach to wax sour'.

Among the most interesting of all scented leaf plants, the Mints are mentioned in *St Matthew's Gospel*: 'Woe unto you . . . for ye pay tithe of mint and anise (Dill) and cummin . . .'

The form *piperascens* is known as the Japanese Mint and during Georgian times the dried and pulverised leaves were imported, to be carried in small silver boxes fastened to the belts of the gentlemen of the time, who would inhale a pinch whenever they desired.

TALL MINT *M. smithiana* Pl. 15
Syn: *M. rubra*
A tall-growing almost glabrous Mint, the leaves almost round, with red veins. The flowers, too, are red with a long hairless calyx and protruding stamens. Closely related to the Whorled Mint, the whole plant has the acrid minty smell of the Corn Mint which figures in its parentage. In bloom June–August.
Habitat: Rare, in ditches and by ponds and streams in S England and Wales.

WATER OR HAIRY MINT *M. aquatica* Pl. 15
Syn: *M. hirsuta*
A perennial growing up to 3 ft tall, its leaves stalked, ovate, serrate and downy on both sides. They release a powerful minty smell when

pressed which is pleasantly refreshing, being almost orange-like. The reddish-purple flowers are borne in axillary and terminal whorls. The calyx is hairy with the stamens protruding. In bloom July–September.

The variety *crispa* has curled leaves, lacerately toothed and with short petioles and is the chief source of oil of spearmint, rather than *M. spicata*, the true Spearmint. The crude oil consists of a terpene and also carvol which has an odour similar to that of Spearmint.

Habitat: Common by the banks of streams and on marshy ground throughout the British Isles.

BERGAMOT MINT   *M. citrata*

A glabrous perennial with rounded leaves often variegated. Possibly a form of *M. aquatica*, it may take on a purple hue though the flowers are more reddish-purple in colour than those of other Mints. The corolla is larger than most and has the stamens inserted inside it. Stalks and calyx are quite smooth and the plant releases the delicious lemony scent of Bergamot (*Monarda*). This is due to the presence of citral as in Lemon Thyme. In bloom July–September.

Habitat: Present in only a few places in England, near the Welsh border of Cheshire (and in N Wales) and in Cambridgeshire and Bedfordshire, beside streams and on marshy ground.

SPEARMINT, LAMB MINT   *M. spicata*
Syn: *M. viridis*

A hairless perennial with an erect stem 12 in. tall and sessile, lanceolate leaves, unequally serrate and brilliant green. They release a warm aromatic odour when pressed, like that of Peppermint though not so pungent. The lilac flowers are borne in attractive spire-like inflorescences, hence its name should be *Spire Mint* not Spearmint. In bloom August–September.

Habitat: Perhaps introduced, but long naturalised by the wayside and in hedgerows as a garden escape.

History: It is the Mint *par excellence* to take with spring lamb while placed in milk it will prevent curdling. Evelyn said, 'spearmint is friendly to a weak stomach and powerful against all nervous crudities'. The distilled water is valuable in relieving hiccuping and flatulence.

PEPPERMINT, BRANDY MINT   *M. piperita*   Pl. 15

A perennial and possibly a natural hybrid between *M. aquatica* and *M. spicata* for it was discovered only after the introduction of the latter species. It grows 12–15 in. tall with stalked, lanceolate leaves serrated at the edges, and the whole plant takes on a purple hue during dry weather. It has a hot, pungent smell, similar to *M. spicata*. There are two forms, the 'black' and the 'white' Peppermint, the former being a coarser plant and more purple, with the upper surface of the leaves purple-brown. Although its oil is more plentiful, it is less refined than that of the 'white' variety which has more deeply serrated leaves. The terminal inflorescence is short and blunt with lanceolate bracts; the calyx tubular; the corolla reddish-purple. In bloom July–September.

Habitat: By the wayside and in ditches, particularly in SE England and in East Anglia; also Lincolnshire and Northamptonshire.

History: It was first discovered in a field in Hertfordshire about the year 1700 and was given its name by Rea in his *Historia Plantarum* (1704) because of its 'hot' peppery smell which is at first cooling and refreshing and then creates a sense of numbness. Turpentine is present throughout the green parts concentrated in the inflorescence just before the flowers begin to open, but peppermint oil owes its value to the solid compound menthol, or methyl alcohol, which has a camphor-like quality. Indeed, the Chinese and Japanese extraction contains this camphor in so concentrated a form that it may occur in a crystalline mass and has been mistaken for true camphor. Menthol, when purified by crystallisation, forms large prisms. It has been used by the Japanese for at least three centuries and at one time was included in the small medicine-box fixed to gentlemen's girdles. The menthol was called Hotan.

In 1879 the *Lancet* drew attention to its medicinal qualities for it was applied to the temples to relieve headaches. It was also used to allay the feeling of sickness and to relieve indigestion. Two drops of oil of peppermint on cottonwool, placed in the hollow of or around a painful tooth, will usually give relief. It is also a valuable antiseptic. 'Oil of peppermint', said Dr Braddon, 'forms the best, safest, and most agreeable of all antiseptics'. Cones made from pure menthol were at one time burnt in sickrooms to ease a hard cough and respiratory difficulty.

Its commercial history dates from 1750 when it was first cultivated at Mitcham in Surrey, where it is still grown to this day. A hundred years later, 500 acres were under cultivation there and at Market Deeping in Lincolnshire, the yield being about 30 lb of oil per acre. Portions of the root were planted in furrows spaced 2 ft apart in a rich, sandy soil. Harvesting for distillation begins early in August and continues, depending upon the weather, for 6–8 weeks during which time the distillation proceeds day and night. The oil extracted improves with age and can be kept for up to 10–15 years. English oil has a purity exceeding all others.

HORSE-MINT, WOODLAND MINT *M. longifolia*
Syn: *M. sylvestris*
A hairy perennial with creeping roots from which the stems rise to a

height of 2 ft with oblong-lanceolate leaves, hairy on the underside and grey-green in appearance. They have a sweet, peppermint smell. The flowers are pale lilac and borne in a slender inflorescence; the bracts are awl-shaped and longer than the flowers; the corolla is hairy on the outside. In bloom August–September.

Habitat: Throughout the British Isles by the wayside; also in ditches and along the banks of streams. It is a handsome border plant to use with Lambsear, *Stachys lanata.*

APPLE OR ROUND-LEAVED MINT
M. rotundifolia   Pl. 15

A perennial, rooting by stolons and bearing erect woolly stems 1–2 ft tall. The leaves are small and oval (almost round) and much wrinkled, with a pleasant smell resembling that of ripe apples. It is distinguished from other Mints by its pale green stems and leaves. The best variety for home use is Bowles Variety. The flowers are lilac or white and are borne in short cylindrical spikes. In bloom August–September.

Habitat: A rare frequenter of deciduous woodlands, chiefly in SW England and S Wales. Occasionally elsewhere in England and Wales.

ORIGANUM   Annual or perennial herbs, their flowers borne in terminal panicles. Bracteoles ovate, imbricate; calyx with 5 equal teeth, hairy in the throat; corolla 2-lipped. Stamens 4, exceeding the corolla. Nutlets free, ovoid, smooth. Mostly Mediterranean though Sweet Marjoram, *O. majorana* was grown in England during Tudor times and was mentioned by Tusser as being suitable for strewing.

MARJORAM   *Origanum vulgare*   Pl. 16

A perennial, it is the only British species. It grows about 12 in. tall and has broad ovate, shortly-stalked leaves covered in down, and flowers borne in a crowded terminal cyme. They are rosy-red with reddish-purple bracts longer than the calyx. In bloom July–September.

Habitat: On dry hilly pastures and banks throughout England and Wales (less common in Scotland and in Ireland), usually growing in a calcareous soil.

History: Like the Sweet Marjoram it has a stimulating scent which is sweet, aromatic and most refreshing during warm weather. Shakespeare held it in high esteem for he frequently mentions it, as in *A Winter's Tale* when Perdita includes it in her list of 'middle summer' flowers (p. 53). From its leaves, and from those of the Sweet Marjoram, 'sweet' waters were made to sprinkle about the home. In Jonson's Masque Chloridia, Rain enters as five persons carrying in their hands

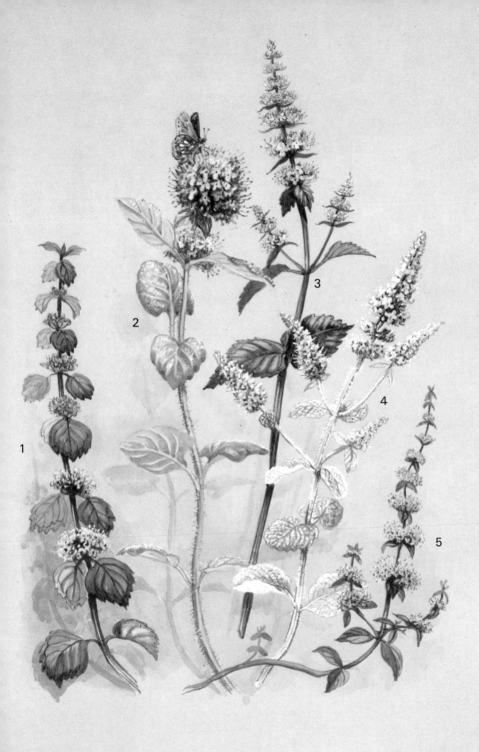

Plate 15
Thyme Family (i)

1 Tall Mint
3 Peppermint
5 Pennyroyal

2 Water Mint
4 Apple Mint

Plate 16
Thyme Family (ii)   1 Wild Basil   2 Marjoram
                        3 Common Calamint 4 Motherwort

pottery balls full of sweet water with which they sprinkle the room, like rain, as they dance.

The dried leaves moistened and placed in a muslin bag brought instant relief when used as a fermentation and applied to rheumatic parts of the body. The Sweet Marjoram secretes an essential oil from the stems and leaves which yields a deposit of crystalline matter known as stearoptene, with a smell similar to camphor.

Parkinson said that the two kinds of Marjoram were much in demand by the ladies 'to put in nosegays' and to use 'in sweet powders, sweet bags and sweet washing waters'. He added, 'our daintiest women do put it still amongst their sweet herbs'. George Withers describes where the plants are to be found in 'A Poet's Home':

> The pleasant way, as up those hills you climb,
> Is strewéd o'er with marjoram and thyme,
> Which grows unset.

In the *Polyolbion*, Michael Drayton wrote of 'Germander, marjoram and thyme' and the plant was described by John Clare in *The Shepherd's Calendar* for June as one of those familiar cottage-garden plants of midsummer:

> With marjoram knots, sweetbrier and ribbon-grass,
> And lavender, the choice of every lass,
> And sprigs of lad's-love, all familiar names
> Which every garden through the village claims.

THYMUS Prostrate or low-growing perennial shrubs, branched and often hairy with small obovate leaves. Flowers pink, purple or white, borne in a terminal spike on an erect shoot. Calyx 2-lipped, throat hairy; corolla 2-lipped, upper lip notched, lower 3-cleft. Stamens 4. Nutlets ovoid, smooth. The flowers secrete an abundance of honey at the base of the ovary. The tube is smooth at the base where the honey is found but is lined with hairs at the top to exclude rain. The flowers are fertilised by Hymenoptera and other insects which, owing to the aggregation of the flowers, are able to pollinate with rapidity.

WILD THYME *Thymus drucei* Pl. 17
Syn: *T. neglectus, T. carniolicus*

A mat-forming perennial undershrub with short 4-angled flowering stems with 2 opposite sides hairy, the other two rarely so. The short-stalked elliptic leaves are borne horizontally and are covered in minute glandular dots. They are pleasantly aromatic when pressed or trodden upon. The reddish-purple 2-lipped flowers are borne in a whorled capitate head and appear earlier than those of most species. In bloom May–August.

Habitat: Throughout the British Isles on dry banks and downlands; also about sand-dunes and on high rocky ground to a height of about 4000 ft.

History: In *Othello*, Shakespeare draws a lesson from this simple plant Wild Thyme. In the Council Chamber in Venice before sailing for Cyprus, the villain Iago addresses Roderigo:

> Virtue? a fig! 'tis in ourselves that we are thus, or thus. Our bodies are our gardens; to the which our wills are gardeners: so that if we will plant nettles, or sow lettuce; set hyssop, and weed up thyme; supply it with one gender of herbs, or distract it with many; either to have it sterile with idleness or manured with industry; why, the power and corrigible authority of this lies in our wills. If the balance of our lives had not one scale of reason to poise another of sensuality, the blood and baseness of our natures would conduct us to most preposterous conclusions . . .

'I like also little heaps, in the nature of mole hills', wrote Bacon of his garden, 'to be set, some with Wild Thyme, some with Pinks, some with Germander'. Moreover, he coupled Thyme with Burnet and Water Mint as those plants which 'best perfume the air'.

'The bees on the bells of thyme', wrote the poet Shelley, for it is the plant most loved by bees. Thyme honey, collected by the bees on Mt Hymettus in Greece and on Mt Hybla in Sicily, is world famous. Spenser spoke of 'the bees alluring Tyme', and possibly because bees are extremely active around the flowers, the plant became an emblem of activity and valour in the days of chivalry. Ingram in his *Flora Symbolica* wrote that ladies 'embroidered their knightly lovers' scarves with the figure of a bee hovering about a sprig of thyme', while 'to smell of thyme' was an expression of praise. Virgil made it the highest compliment the shepherd could pay to his mistress:

> Nerine Galatea, Thymo mihi dulcior Hybla.

In *Britannia's Pastorals*, William Browne wrote:

> Some from the fens bring reeds, wild thyme from downs.

As honey was relied upon for sweetening in medieval times, bees were highly valued and the plants they visited were therefore considered of greatest importance. In *The Parliament of Bees*, a charming little play written by John Daye in 1641, all but one of the characters are bees. The Master Bee proclaims the freedom of gardens and meadows:

Of wanton Cowslips, Daisies in their prime
Sun-loving Marigolds, the blossom'd Thyme,
The blew vein'd Violets, and the Damask Rose,
The statelie Lily, mistress of alle those . . .

'The owners of hives have a perfite forsight and knowledge what the increase or yields of honey will be every year by the plentiful or small number of flowers growing and appearing on the thyme about the Summer solstice', wrote Thomas Hill in his *Gardener's Labyrinth* (1577).

And Parkinson in the *Paradisus* wrote, 'There is no herb of more use in the houses of high and low . . . for bathing, for stewings and to make sauces for fish and flesh. We preserve them with all the care we can in our gardens for the sweete and pleasant scents and varieties they yield'.

LARGER WILD THYME *T. pulegioides*

Syn: *T. chamaedrys*

A tufted perennial of low, upright habit with ascending square stems, the angles covered in short hairs. The hairless elliptic leaves are larger than those of *T. drucei* and release a powerful aromatic smell when pressed. The rosy-purple flowers are borne in small axillary heads on branches which ascend from the crown of the rootstock. There is a short and broad upper lip to the corolla. In bloom June–September.

Habitat: On dry grassy banks and downlands, usually in calcareous soils and mostly confined to E England, especially Kent, Suffolk and NE Yorkshire; also central Ireland, particularly Cavan, Longford and Roscommon.

History: It is grown in S France for its essential oil which is reddish-brown in colour and highly fragrant. It is separated into two parts by fractional distillation. The first part is a mixture of cymene and thymene; the second is thymol, a powerful antiseptic which is about ten times as powerful as carbolic acid but is at the same time non-poisonous and agreeably scented.

Crossed with *T. vulgaris*, *T. pulegioides* has produced the Lemon Thyme, *T. citriodorus*, with a refreshing lemon smell due to a combination of citral and eucalyptol. The variety Silver Queen has white variegated leaves with the rich lemon scent.

MOUNTAIN THYME *T. serpyllum*

An almost prostrate perennial, it differs from *T. drucei* only in its stems which are completely covered with evenly distributed white hairs. The leaves are borne upright and are egg-shaped, broader at the tips, while the flowers, with their deep red calyx and rose-coloured corolla, are borne in terminal heads. In bloom June–August.

Habitat: Only on the Brecklands of East Anglia, growing in sandy heaths and in grasslands.

History: In an Anglo-Saxon book of the tenth century the name *Serpulum* appears and in the *Promptorium Parvulorum* (1440), the plant is referred to as *Serpillum piretrum*. The name *thyme* or *thymum* first appeared in the Catholic *Anglicum* since when the name has continued in use. It is probably derived from the Greek *thumos*, sacrificial smoke, a reference to the ancient use of the dried stems and leaves in sacrifices because of its pleasant odour, and later its use in places of worship.

In ancient Greece, the plant was considered to be above all others in its medicinal qualities. It is anti-spasmodic, good for nervous headaches and was considered an excellent restorative against tiredness, as Virgil mentioned in the *Eclogue*, translated as follows:

> Thestlis for mowers tired with parching heat,
> Garlic and Thyme, strong smelling herbs, doth beat.

Countryfolk used to make a 'tea' from an infusion of the leaves, which had a sweet and fragrant odour, while Gerard said that 'boiled in wine and drunk, it is good against the warblings and rumblings of the belly'. Oil of thyme is recommended to this day to arrest gastric fermentation. When a few drops are placed in boiling water and inhaled, the oil will relieve a stuffy feeling caused by a cold in the head.

A number of varieties make delightful garden plants to set between paving stones or on a dry wall, for they are neat of habit, long in bloom, and most colourful. The variety *coccineus* bears mats of brilliant crimson, while Pink Chintz and the white-flowered *albus* are also suitable for garden planting.

CALAMINTHA Perennial herbs with a creeping rootstock bearing erect stems clothed in hairs. Leaves ovate, rounded at the base and serrate at the edges, hairy on both surfaces. Flowers borne at leaf-axils or in loose terminal panicles; calyx tubular, 2-lipped, hairless; corolla 2-lipped, upper lip flat, lower spreading. Stamens 4, shorter than corolla and converging at tips. Nutlets smooth.

COMMON CALAMINT *Calamintha ascendens* Pl. 16
Syn: *C. officinalis*
A perennial shrublet growing 12–15 in. tall with a creeping rootstock and stems covered in downy hairs. The leaves are ovate, serrated at the edges with long petioles. The flowers are lilac-pink and borne in forked axillary cymes, with pointed bracts. The calyx is 2-lipped with long hairs on the teeth. The lower lip of the corolla is lobed and covered with dark purple spots. In bloom July–September.

Plate 17

Thyme   1 Wild Thyme   2 White Horehound
Family (iii)   3 Hedge Woundwort   4 Black Horehound

Plate 18
Thyme
Family (iv)

1 Ground-pine
3 Bastard Balm

2 Balm

Habitat: On dry banks of a calcareous nature throughout England and Wales, also W Ireland and the Isle of Man. Rare in Scotland.

History: It takes its name from the Greek *kalos*, beautiful, and *minthe*, mint, referring to the beauty of the plant and the aromatic minty smell of its leaves.

LESSER CALAMINT   *C. nepeta*
Syn: *C. parviflora*

A perennial with erect stems growing 15–18 in. tall which are more branched than those of *C. ascendens*. The ovate leaves are short-stalked and serrate, with hairs on the underside. Their minty scent is not so pronounced as in *C. ascendens*. The pale purple flowers are borne in forked cymes, the calyx has long hairs on the teeth; the corolla is free from spots on the lower lip. In bloom July–September.

Habitat: On dry banks and downlands in calcareous soils but mostly confined to Kent and Sussex; Suffolk; E Yorkshire and S Wales.

CLINOPODIUM Annual or perennial herbs. Leaves ovate, slightly toothed. Flowers pink, borne in axillary or terminal whorls; bracts and calyx hairy. Corolla-tube curved distinguishing it from *Calamintha*; it is usually smooth within, the upper lip flat, the lower spreading. Stamens 4. Nutlets smooth.

WILD OR BUSH BASIL   *Clinopodium vulgare*   Pl. 16
An aromatic straggling perennial growing 1–2 ft tall and with a creeping rhizomatous rootstock. The slightly toothed ovate leaves are stalked and are deeply veined while the rose-pink flowers are borne in crowded axillary whorls. They are furnished with long bristly bracts, white with hairs and are visited mostly by butterflies. In bloom July–September.

Habitat: On dry banks and grassy hillsides of a calcareous nature; also in hedgerows and along the borders of woodlands, mostly in England south of a line drawn from the Mersey to the Humber. Rare elsewhere.

History: It takes its name *Clinopodium* from the Greek meaning 'a foot-stool', a reference to the hairy bracts which form a 'stool' upon which the pink flowers 'sit'. The Wild Basil was one of the plants mentioned by Parkinson as being used in nosegays, together with Sweet Marjoram, Maudeline and Costmary. It was the 'tufted basil' of Shenstone's poem 'The Schoolmistress':

> Herbs too, she knew, and well of each could speak
> That in her garden sipp'd the silv'ry dew; . . .
> The tufted basil, pun-provoking thyme,
> Fresh balm, and mary-gold of cheerful hue.

Tusser included it in his list of herbs used for strewing, together with twenty others including Balm and Costmary, Germander and Penny-royal, which were to be gathered fresh and would release a powerful refreshing smell when trodden upon.

MELISSA Perennial herbs with erect hairy stems. Leaves ovate, crenate. Flowers white, borne in one-sided axillary whorls; calyx 2-lipped; corolla 2-lipped, tube curving upwards. Stamens 4, shorter than corolla. Nutlets smooth.

BALM *Melissa officinalis* Pl. 18

A perennial bearing an upright hairy stem 1–2 ft tall with stalked ovate leaves, pale green in colour, deeply wrinkled and serrate. The flowers are blush-white, borne in short-stalked axillary whorls from the upper part of the stem and leaf-joints. The calyx is covered in long white hairs. The leafy parts of the plant are deliciously scented when bruised, more so than any other Labiate. In bloom July–September.

Habitat: It may have been introduced by the Romans for it has been long naturalised or may be indigenous to England. As Gerard said, is found in woodlands but more often on mountainous slopes, possibly as an escape from cottage gardens.

History: Its leaves are lemon-scented and contain citral, present also in the leaves of the Lemon-scented Geranium and in Lemon Thyme; also in Lemon-scented Verbena, which has the most appreciated of all scented leaves and was at one time much in demand for pot-pourris. 'Our common Bawne', wrote Gerard, 'having many square stalks and blackish leaves, of a pleasant smell, drawing near in smell and savour unto Citron . . .'. 'Balm', wrote Evelyn, 'is sovereign for the brain, strengthening the memory and powerfully chasing away melancholy'. The leaves were placed in ale and wine to impart their flavour and were thought to be a restorative, in the same way that Balm 'tea' was drunk in the cottager's home. Together with nutmeg, lemon peel and the root of Angelica, spirit of balm was the principal ingredient of Carmelite Water. It was considered a valuable restorative. Virgil in the *Georgics* wrote:

> . . . liquors cast in fitting sort,
> of bruised Bawne and more base Honeywort.

It is not to be confused with Balm of Gilead, obtained from a gum-bearing plant of the Near East, which from earliest times has been used to heal wounds. Shakespeare refers to the 'healing balm' on twenty occasions but only two represent the native Sweet Balm. It was mentioned in the famous last words of Cleopatra as she applies an asp to her arms and falls dying from its poison:

As sweet as Balm, as soft as air, as gentle
O Antony! Nay I will take thee too . . .

The sweetness of Balm was proverbial during Elizabethan times, for besides its medicinal properties it was used to make chaplets and garlands. In the *Muses' Elysium*, Michael Drayton wrote that 'balm and mint help to make up my chaplet'. For its sweet fragrance it was included among Tusser's twenty-one 'strewing herbs' (p. 20) and Michael Drayton mentioned it, too, in the *Polyolbion*, coupling it with the Mint.

In Tudor times, the juice was extracted from the leaves and stems to make wine, and it was also used to rub on furniture to impart its lemony scent. Shakespeare alludes to this in *The Merry Wives of Windsor*, probably written in 1600. In Act V, Scene V, Falstaff, wearing a buck's head, has entered Windsor Great Park, just as the clock has struck twelve. He is joined by Sir Hugh Evans, dressed like a satyr, by Mistress Quickly and Pistol, and by Anne Page as the Fairy.

> Fairy Queen: About, about:
> Search Windsor Castle, elves, within and out:
> Strew good luck, ouphes, on every sacred room;
> That it may stand till the perpetual doom,
> In state as wholesome, as in state 'tis fit,
> Worthy the owner, and the owner it.
> The several chairs of order look you scour
> With juice of balm and every precious flower
> Each fair instalment, coat, and several crest,
> With loyal blazon evermore be bless'd!

Its botanical name *Melissa* is a reference to the large secretion of honey in the flowers, while *Balm* is an abbreviation of *Balsam* which it resembles in its scent. Its juice was used to heal wounds in the same way. Dioscorides said that the leaves 'being applied, do close up wounds without any perill of inflammation', and Gerard said that the leaves would 'heal up green wounds made with irons (swords)'. The leaves, with those of Lemon Thyme, Lavender, Rosemary and Bay may be used in pot-pourris, for they retain their lemony scent after drying.

To make a moist pot-pourri, gather 2 lb of red rose-petals (those of the Ena Harkness, Fragrant Cloud and Wendy Cussons varieties are the most fragrant) and cover with $\frac{1}{2}$ lb of common salt. Allow to stand for 7 days, then add to them $\frac{1}{2}$ lb of bay salt and the same amount of ground cloves and brown sugar, $\frac{1}{4}$ lb of gum benzoin, 2 oz of powdered orris root, a cupful of brandy and the dried leaves of Lavender, Balm, Rosemary and Verbena or Lemon Thyme. The mixture should be kept in a covered jar and should be frequently stirred. The brandy will long preserve the scent.

SALVIA   Annual or perennial herbs or undershrubs growing 1–3 ft tall. Leaves roughly toothed, wrinkled, grey-(sage) green in colour, with a pungent smell. Flowers borne in axillary whorls; calyx and corolla 2-lipped; stamens 2, branched. Nutlets ovoid, smooth.

MEADOW SAGE   *Salvia pratensis*

A perennial growing about 2 ft tall with hairy stems and glandular, pubescent, narrow leaves which release an aromatic smell when handled. The flowers are either hermaphrodite or female, on different plants, and are borne in handsome spikes of brilliant blue. The calyx is downy but without long hairs. The bracts are shorter than the calyces. The corolla-tube is long with a hooded upper lip, glandular on the outside. Pollination is by the longer-tongued bees. In bloom June–July.

Habitat: Meadow Sage is found on grassy downlands of S England from Kent and Sussex to Wiltshire; also NE Lincolnshire and E Yorkshire wherever there is a calcareous soil.

History: The ancients gave pride of place over all garden plants to the Sages. In his ninth-century poem, Walafrid Strabo wrote, 'Amongst my herbs, sage holds place of honour; of good scent it is and full of virtue for many ills'. The plant was named *Sage*, meaning salvation, for its ability to cure 'many ills'; it was also 'full of virtue, wise, mellow'. The word was frequently used by Shakespeare in this context, though he made no direct reference to the plant in his plays. In *Richard III*, in the courtyard at Baynard's Castle, the Earl of Gloucester addresses the Lord Mayor, Aldermen and Citizens of London as 'Cousin of Buckingham and sage grown men'. Again in *Henry IV, Part 2*, Shakespeare uses the term 'sage councellors'.

Sage is reputed to give long life and has been used through the years for curing all manner of complaints. Sage cordial, best made from the Red Leaf variety, is most valuable for a sore throat, while an infusion of its dried leaves (an ounce to a pint of boiling water) taken when cold, is an excellent aid to digestion as well as a help to those who suffer from anaemia. It may also be used (with Rosemary) as a hair tonic, and prevents the hair falling out better than any other preparation, while it will also darken greying hair. Sage water should be massaged into the scalp every day after rising, with a little brilliantine once a week to prevent the hair from becoming too dry. The hair will quickly respond to this treatment.

From earliest times, Sage has been used in stuffing for the richer meats and game, and more dried Sage is sold for this purpose than any other herb. But it is important to obtain a broad leaf to use for stuffing, though for medicinal purposes the narrow leaf form is most desirable. The active principle of the essential oil is salviol which has the power to resist putrefaction of animals, hence its use with meats, while its bitter pungency enables the stomach to digest cooked meats more easily. Spenser wrote, 'The wholesome sage and lavender still grey'.

In Russell's *Boke of Nurture* there is a description of sage fritters served at banquets in the Middle Ages, and the seeds were used to flavour cheeses, a custom handed down from the time of the Romans. Evelyn suggested using the leaves 'but principally the flowers in our sallets'.

Turner's *Great Herbal* contains a remedy for forgetfulness which consists of making a decoction of 'tutsan, smalage and sage' and bathing the back of the neck and head with it. 'Sage', wrote Gerard, 'is singular good for the head and brain; it quickeneth the senses'.

CLARY *S. horminoides*

A glandular perennial growing about 12 in. tall with square stems and hairy ovate dark green leaves, serrated at the edges. The stems are tinted with purple at the top. The flowers are deep violet-blue with 2 white marks on the lower lip. The calyx is glandular with long white hairs at the base. In bloom June–September.

Habitat: Gerard said that it 'groweth wild in divers barren places, especially in the fields of Holborn, neare unto Grayes Inne . . . at the end of Chelsey next to London; and in the highway as you go from the Queen's Palace of Richmond to the water's side'. In other words, on wasteland and in meadows mostly in England, south of the Thames. Rare elsewhere.

Clary

History: Its name *Clary* is a shortened version of *clear-eyes*, as from earliest times an infusion of the aromatic leaves or the seeds has been used to treat tired or inflamed eyes. For this reason the plant was known as *Oculus Christi*. Its leaves emit a refreshing pineapple scent and in Europe they were used in the manufacture of perfumes and to impart their delicious fruity scent to jellies. In Tudor times, they were dipped in batter and used in omelettes and were in demand for mixing in pot-pourris and filling scent bags. The French made a wine, remarkable for its narcotic qualities, from the leaves, and these were also used in ale to make it more potent and to impart their refreshing flavour.

The leaves of the garden variety *Salvia sclarea*, which are large and thick and covered in hairs, have a rather unpleasant fur-like smell but on distillation yield an oil with a pleasant muscat scent.

BASTARD BALM   *Melittis melissophyllum*   Pl. 18
The only species of the genus, it is a hairy perennial growing 1–2 ft tall with hairy stems and large obovate leaves, serrated at the edges. The handsome flowers are large, the largest of all the Labiates, and are white, spotted with pink on the lower lip. They are pollinated mostly by night moths. There is a form *grandiflora* which has a red middle lobe on the lower lip. The tube is long so that the honey is available only to hawkmoths. The whole plant emits a powerful pungent smell of sage-like quality. In bloom May–August.
Habitat: At the margins of woodlands and on dry banks but almost entirely confined to SW England and Wales, and especially common in Pembrokeshire, Gloucestershire, Somerset and Cornwall.

STACHYS   Hairy annual or perennial herbs with ovate or lanceolate leaves, serrate and usually with long petioles. Flowers borne in terminal spikes; calyx tubular with 5 narrow teeth. Corolla-tube as long as calyx with ring of hairs inside; upper lip arched, lower 3-lobed. Nutlets rounded at one end. Pollination by bees, but with several species self-pollination is possible.

MARSH WOUNDWORT   *Stachys palustris*
A hairy perennial with an erect hollow stem 2 ft tall and linear-lanceolate leaves, serrated at the edges, the lower short-stalked. The flowers are borne in a terminal spike of 6–8 whorls and are of dull pale red with a long tubular calyx, sometimes purple in colour. The leaves have a faint but refreshing smell similar to that of *Rosa rubiginosa*. In bloom July onwards.
Habitat: Common throughout the British Isles in hedgerows and on the banks of lakes and streams.

HEDGE WOUNDWORT   *S. sylvatica*   Pl. 17
A hairy perennial resembling *S. palustris* but with solid stems growing 3 ft tall. The coarse leaves are cordate, stalked and serrate and emit a pungent and somewhat unpleasant smell when handled. The flowers, which resemble small Antirrhinums, are borne in a whorled spike and are crimson-purple blotched with white. The calyx is hairy and glandular with teeth about half as long as the tube. In bloom July–August.
Habitat: Common in hedgerows and woodlands throughout the British Isles, especially SE England.
History: Gerard named it *Clown's All-heal* and tells how a farm-

labourer cut his leg to the bone when using a scythe and healed it within a week by applying the plant's leaves. 'The leaves, pounded with hog's grease and applied to green (new) wounds in the manner of a poultice, heal them in such short time that it is hard for anyone who has not had the experience thereof to believe'.

It takes its name *Stachys* from the Greek *stachus*, a bunch, from the arrangement of the flowers. The plant contains a volatile oil with antiseptic qualities.

Gerard said that the plant grew 'in Kent about Southfleet, near to Gravesend, and likewise in the meadows by Lambeth, near to London'.

BALLOTA Hairy perennials often of unpleasant smell. Leaves cordate, downy, wrinkled. Flowers borne in axillary whorls with leafy bracts; calyx funnel-like with 5 spreading teeth; corolla 2-lipped, upper lip erect; lower 3-lobed. Stamens 4; anthers exserted. Nutlets oblong.

BLACK HOREHOUND *Ballota nigra* Pl. 17

A much-branched hairy perennial of unpleasant smell, growing 1–2 ft tall with cordate leaves, stalked and serrated. The flowers are pale purple-red and are borne from the axils of the leaves all the way up the stem. The calyx is funnel-shaped, the tube being shorter with a circle of inner hairs. The whole plant gives off an unpleasant rancid smell when approached, and if handled the smell is difficult to remove from the fingers. In bloom June onwards.

Habitat: Common in hedgerows and on waste ground throughout England and Wales; rare in Scotland and Ireland.

History: It takes its name from the Greek, *ballo*, to reject, a reference to its offensive smell, though Meyrick said, 'this is one of those neglected English herbs which are possessed of great virtues, though they are little known'. The leaves, applied externally, brought relief to sufferers from gout and were supposed to act as an antidote to the bite of a mad dog, a fact mentioned in Beaumont and Fletcher's *The Faithful Shepherdess*:

> Black Horehound, good
> For sheep, or shepherd bitten by a mad-dog's venomed teeth.

YELLOW ARCHANGEL *Galeobdolon luteum* Pl. 19

A genus of a single species, and distinguished from *Lamium* by its more slender appearance and smaller number of hairs. It has stalked ovate leaves, rounded at the base and coarsely serrated – 'very much cut or hackt about the edges', wrote Gerard – while the handsome flowers are borne in axillary whorls. They are bright yellow with red spots on the lower lip of the 2-lipped corolla. They act as guides towards the honey contained in the long hairy tube which is longer than the calyx. Pollination is by bees. In bloom May–July.

Habitat: Mostly in thinned woodlands and hedgerows of England and Wales. Rare in Scotland, and in Ireland where it is present only on the eastern coastline.

History: Gerard wrote, 'that with the yellow flowers groweth not so common as the others. I have found it under the hedge on the left hand as you go from the village of Hampstead, near London to the church . . . and in the woods belonging to the Earl of Cobham in Kent'. He added, 'the flowers are baked with sugar as roses are . . . as also the distilled water of them which is used to make the heart merry and . . . to make a good colour in the face'.

LAMIUM   Hairy annual or perennial herbs distinguished from Stinging Nettle by their square stems, whilst flowers of the latter are green. Leaves ovate, cordate at the base, serrate and with petioles. Flowers borne in whorls from axils of leafy bracts; calyx tubular or bell-shaped; 5-toothed; corolla with arched upper lip, lower 2- or 3-lobed. Stamens 4; anthers hairy. Pollination mostly by Hymenoptera, or self-pollinating. It takes its name from the Greek *laimos*, the throat, from the shape of its corolla.

RED DEAD-NETTLE   *Lamium purpureum*   Pl. 19

A hairy annual growing 6–12 in. tall with oval serrated leaves tinted purple but leafless near the base of the flower stems. The wrinkled leaves and stems give off a somewhat aromatic pungent smell when bruised. The flowers, borne in axillary whorls, are pale purple-red with a hooded upper lip and the lower lip 2-lobed. The tube is longer than the calyx, with an internal ring of hairs. In bloom May–October.

Habitat: By the wayside and at the edges of woodlands throughout the British Isles.

WHITE DEAD-NETTLE   *L. album*

A hairy perennial of creeping habit, growing less than 6 in. tall with ovate stalked leaves deeply serrated and blotched with white. The flowers are white, remaining wide open, and are borne in axillary whorls; the thin calyx-teeth as long as the tube, with the corolla-tube longer than the calyx. The handsome flowers have contrasting black anthers. The plant is sweetly aromatic when handled. In bloom March–October.

Habitat: Common in hedgerows and on waste ground throughout England and Wales. Rare in Scotland and Ireland.

3

2

1

4

Plate 19
Thyme Family (v)   1 Ground Ivy   2 Yellow Archangel
                          3 Wild Catmint   4 Red Dead-nettle

Plate 20
Honeysuckle 1 Black Elder 2 Honeysuckle
Family (i)

LEONURUS Perennial herbs. Leaves radical with large deep lobes. Flowers white or pink, borne in axillary whorls; calyx bell-shaped with 5 prickly spreading teeth; corolla with flat upper lip; anthers covered with spots. Pollinated by Hymenoptera.

MOTHERWORT *Leonurus cardiaca* Pl. 16
A hairy perennial with unbranched leafy stems growing 2–3 ft tall, its radical leaves palmately 3–5-lobed and held on long stalks. The pink flowers spotted with mauve are borne in whorls; the calyx furnished with 5 prickly teeth. The plant emits an aromatic pungent smell when bruised. In bloom July–September.
Habitat: On wasteland usually near old buildings. Scattered throughout the British Isles but rare.

NEPETA Perennial herbs with branched hoary stems. Leaves ovate or cordate, coarsely serrate and tomentose. Flowers borne in axillary whorls forming terminal inflorescence. Calyx tubular, 5-toothed; corolla-tube glabrous, longer than calyx. Upper lip flat, notched; lower lip, 3-lobed. Nutlets ovoid.

WILD CATMINT *Nepeta cataria* Pl. 19
An erect hoary perennial growing 1–2 ft tall with cordate leaves, downy on the underside and strongly aromatic. The flowers are borne in crowded axillary whorls and are white, flecked with red. The calyx is pubescent and 5-toothed. Pollinated by bees and butterflies. In bloom July–September.
Habitat: Throughout England, especially in calcareous soils, in hedge-rows and on waste ground. Rare in Scotland and in Ireland where it is possibly an introduction.
History: Gerard called it *Herba Cattaria* 'because cats are very much delighted herewith; for the smell of it is so pleasant unto them that they rub themselves upon it and wallow and tumble in it, and also feed on the branches and leaves very greedily'. Gerard said that 'the whole herb is soft and covered with a white down'. Abraham Cowley wrote:

> Lavender, Corn-rose, Pennyroyale sate,
> And that which cat's esteem so delicate.

All parts of the plant have a pungent smell, lightened with a lemony scent. The root is especially aromatic and was said to give one courage, hence the legend that the dried root was chewed by hangmen when carrying out their gruesome occupation in public.

GLECHOMA Perennial herbs with creeping stems. Leaves ovate, cordate at base. Flowers purple-blue, borne in small, loose axillary whorls. Calyx tubular, 2-lipped, hairy at base of lower lip. Tube narrow, straight. Nutlets obovoid.

GROUND IVY, ALE-HOOF   *Glechoma hederacea*   Pl. 19
Syn: *Nepeta glechoma, N. hederacea*

An almost glabrous perennial with creeping stems, rooting at the nodes. The leaves are heart-shaped, bluntly toothed and sage-green in colour, irregularly marked and edged with purplish-white. They have long petioles and are pleasantly aromatic. The flowers are purple-blue, borne 3 or 4 at the leaf-axils. The lower lip is spotted with dark purple. In bloom April–July.

Habitat: Throughout the British Isles, it is usually present in woodlands and beneath hedgerows, growing in partial shade.

History: With its heart-shaped leaves and trailing stems it is a delightful plant, especially the more refined garden form, when used in window boxes and hanging baskets, as it retains its variegated leaves through winter. When on the ground, the plant sends out its wiry stems to a distance of several yards, rooting at the leaf-nodes. Pieces may then be detached and planted in small pots for growing on. 'Amongst the leaves', wrote Gerard, 'come forth the flowers, gaping like little hoods, not unlike those of Germander'. Gerard said that the women of the North, especially of Wales and Cheshire (Gerard was a Nantwich man) 'do turn the herb Ale-hoof into their ale' which it clarified and to which it imparted its aromatic balsam-like smell. For that reason, it was always to be found growing in the gardens of wayside inns. It was also grown in every cottage garden, for the leaves were made into 'tea' to purify the blood, and when dried were used as snuff to clear a cold in the head or relieve headaches. It was one of the plants sold in the streets of London as the *Roxburghe Ballads* mention:

> Here's fine rosemary, sage and thyme.
> Come, buy my Ground Ivy.
> Here's featherfew, gilliflowers and rue.
> Come, buy my knotted marjoram, too!

Under a microscope the essential oil may be seen exuding from glandular dots on the under surface of the leaves and is released when the leaves are pressed. The plant took its earlier name from Nepet in Tuscany where it abounds, while *glechon* is Greek for mint.

MARRUBIUM Hairy perennial herbs with branched square stems. Leaves ovate, cordate at the base. Flowers usually white, borne in axillary whorls; calyx tubular with 5–10 prickly teeth, hairy in throat. Corolla-tube larger than calyx, two-lipped; upper 2-cleft, lower 3-lobed. Nutlets ovoid.

WHITE HOREHOUND   *Marrubium vulgare*   Pl. 17

A bushy aromatic perennial growing 1–2 ft tall, it is covered in woolly
down which gives it a frosted appearance. The oval, bluntly-toothed
leaves are heavily wrinkled, the lower having long stalks. The tiny
white flowers are borne in dense whorls from the leafy bracts and have
10 sharply hooked calyx-teeth. In bloom June–September.

Habitat: Widespread throughout England and Wales on downlands
and on waste ground by the wayside, usually where the soil is of a
calcareous nature. Rare in Scotland and Ireland.

History: The plant takes its name from the Hebrew word *marrob*
meaning bitter juice, and is distributed throughout S Europe and the
Near East. It is reputed to be one of the five bitter herbs of the *Mishna*
which the Jews were ordered to take during the Feast of the Passover.
In medieval times, syrup of horehound and candied horehound were
acknowledged remedies for coughs. To candy, one should boil the
fresh leaves and add sugar, then re-boil the mixture and allow it to cool.
Michael Drayton wrote in the *Polyolbion*: 'Pale hore-hound, which he
holds of most especiall use'. And again in the *Muses' Elysium* he speaks
of its use as a cure for the bite of a mad dog (see Black Horehound,
p. 171):

> Here hore-hound 'gainst the mad dog's ill
> By biting, never failing.

A delicious beer was made from the stems and leaves and it was
grown for this purpose in every East Anglian cottage garden. The plant
was once sold in the streets of London, for it had numerous uses:

> Here's dragon's tongue and wood sorrel,
> With bear's-foot and horehound.
> Let none despise the merry, merry cries
> Of famous London Town.

The leaves, soaked in water and left exposed to the sun, produce a
liquid of unpleasant smell but which, when distilled and dissolved in
ether, has a sweet aromatic odour and a burning taste. Walafrid Strabo
described it as 'bitter to the palate, yet its scent is sweet'. When warmed
by the sun, the leaves have a musky aroma about them which Miss
Sinclair Rohde described as being of 'a curious dusty Eastern fragrance'.

TEUCRIUM   Perennial herbs often with woody stems and rhizomatous rootstock. Leaves oblong or sessile, coarsely serrate. Flowers
usually borne in axillary whorls, the upper in a loose terminal raceme.
Calyx tubular, 5-toothed; corolla with short tube. Upper lip 2-cleft or
absent; lower lip 3-cleft. Stamens 4. Nutlets smooth or wrinkled.

WALL GERMANDER   *Teucrium chamaedrys*
A branched woody perennial with ovate leaves, coarsely serrate and of
darkest green, shiny above, hairy on the underside. The flowers are
large and handsome, rosy-purple in colour, the upper borne in a
terminal 1-sided raceme. Pollination is by bees but self-pollination is
also possible. In bloom June–September.
Habitat: Possibly introduced by the Normans, it is to be found on old
castle and monastery walls in England and Wales. Rare in Scotland and
Ireland.
History: It was included among Tusser's twenty-one herbs for strewing
(p. 20) and it was used for this purpose until Stuart times. Parkinson,
writing of the plants suitable for making 'knotted' beds, said of Germander and Hyssop, 'they must be kept in some form and proportion
by cutting, and the cuttings are much used as strewing herbs for houses,
being pretty and sweet, with a refreshing lemony scent'. The Elizabethan
Sir Hugh Platt, 'knight of Lincoln's Inn', who had a famous garden
nearby, suggested several plants to grow indoors. He also had an estate
near St Albans and later owned Kirby Hall in Northamptonshire,
where his garden still survives and remains much as it was during his
lifetime. He advised using pots of Rosemary, Sweetbriar, Bay and
Germander for the more shady parts of a room, and Dr Leminius, a
Dutch physician, when on a visit to England in 1560, wrote of the
Englishman's home, 'their chambers and parlours strewed over with
sweet herbs refreshed me, their nosegays finely intermingled with
sundry sorts of fragrant flowers, in their bed chambers and privi rooms
with comfortable smell, cheered me up and entirely delighted my
senses'. The leaves may be dried and used in pot-pourris, to which they
impart a scent like that of Balm and Lemon-scented Thyme.

AJUGA Annual or perennial herbs with leafy bracts. Flowers
small, borne in axillary whorls to form a terminal inflorescence. Calyx
tubular, 5-toothed; corolla with ring of hairs within tube. Upper lip
short; lower 3-lobed. Nutlets wrinkled.

GROUND-PINE   *Ajuga chamaepitys*   Pl. 18
A tufted hairy annual with spreading reddish-brown stems and almost
prostrate habit. The hairy grey-green leaves are divided to the base into
3 linear lobes which, due to the presence of turpentine, release a
resinous smell like pine leaves when trodden upon. The bugle-shaped
flowers, borne in pairs from the leaf-axils along the stem, are yellow
with red spots in the throat. In bloom June–September.
Habitat: On arable land and sand-dunes in SE England and East
Anglia. Rare elsewhere.

History: Culpeper said that the plant smelled 'somewhat like unto strong resin', while a decoction of the leaves, mixed with the pulp of figs, was taken as a gentle laxative. It was also used as a remedy for coughs. Abraham Cowley wrote, 'Ground pine, with her short legs, crept hither too'.

PLANTAIN FAMILY   Plantaginaceae

Annual or perennial herbs mostly of the northern temperate regions with radical leaves, often ribbed and spirally arranged. Flowers hermaphrodite, borne in spikes or racemes. Calyx 4-parted, persistent; corolla 4-parted, imbricate. Stamens 4, thread-like, alternating with corolla segments. Filaments long, anthers broad; ovary usually 1–2-chambered; style 1, stigma hairy. Fruit, a capsule. Almost all species are wind-pollinated and scentless, *Plantago media* being the exception for it is also insect-pollinated.

PLANTAGO   Perennial herbs, glabrous or pubescent with ribbed leaves of rosette formation which lie flat on the ground. Flowers borne in cylindrical spikes. Calyx 4-cleft; corolla tubular with 4 lobes. Stamens 4, inserted in tube and extending beyond. Ovary 2–4-chambered.

HOARY PLANTAIN
Plantago media
A pubescent grey plant with elliptic leaves, prominently ribbed and formed in a rosette. The flowers are borne on 6–10-in. stems in dense cylindrical spikes 2–3 in. long. They are purple-pink in appearance, as this is the colour of the conspicuous filaments. The plant is visited by bees and is deliciously scented, but the cream-coloured anthers are so large and the pollen so smooth and abundant that wind-pollination also takes place. In bloom May–August.

Hoary Plantain

Habitat: In grasslands and hedgerows throughout England but most common in the south and midlands. Rare in the north and in Scotland and Ireland.
History: It takes its name from the Latin *planta,* the sole of the foot, an allusion to the flat leaves which lie on the soil. Countrymen applied

the leaves to leg-wounds and Shakespeare referred to this in *Romeo and Juliet* when Romeo says, 'your plantain leaf is excellent for broken shin'. The root is slightly scented and has a sweet taste, while the leaves if macerated and the distillate treated with ether, yield a ferment oil of a sweet, slightly aromatic scent resembling musk.

BEDSTRAW FAMILY Rubiaceae

Trees, shrubs or herbs. Leaves opposite or whorled with stipules between the petioles and often united around the stem. Flowers minute, borne in terminal or axillary cymes, highly scented in *Chinchona* and *Coffea* of tropical and warm temperate regions. British species herbaceous with angular stems, belonging to tribe Galieae, the Bed-straws. Flowers polysymetric, 4–6-numerous; sepals free, usually indis-tinguishable; corolla tubular. Ovary 2-chambered; fruit a berry or capsule, used in making a beverage (coffee). Flowers and leaves of a number of species have scented attractions, e.g. *Asperula cynanchica* which has vanilla-scented flowers and is visited by bees and flies; the leaves of several species contain coumarin and smell of newly mown hay.

ASPERULA Perennial herbs, usually glabrous with leaves borne in whorls. Flowers in terminal panicle or corymb; calyx with 4–5 teeth; corolla funnel- or bell-shaped; white, pink or blue, 4-lobed. Stamens alternating with petals; anthers purple to attract flies. Fruit, 2 single-seeded mericarps.

SQUINANCY WORT *Asperula cynanchica*
A perennial herb of spreading, prostrate habit, the stems 4-angled and glabrous. The leaves are obovate, often unequal and borne 4 in a whorl, while the white flowers, pink on the outside, are borne in long-stalked terminal cymes. They have a powerful vanilla-like scent which is especially pronounced when warmed by the sun. In bloom June–August. *Habitat:* On calcareous mountainous pastures, chiefly in E Yorkshire and in Ireland.

GALIUM Annual or perennial herbs bearing their leaves and stipules in whorls. Flowers hermaphrodite or polygamous, borne in terminal or axillary cymes, differing from *Rubia* in the rotate, 4-lobed corolla and from *Asperula* in its short tube. Stamens 4; styles 2; stigmas capitate. Fruit composed of 2, 1-seeded mericarps, glabrous or hairy.

WOODRUFF *Galium odoratum*
Syn: *Asperula odorata*
An almost glabrous prostrate perennial with erect 4-angled stems

6–8 in. tall, hairy beneath the nodes. The lanceolate leaves are produced in whorls of 6–9 with prickles at the margins. The chalk-white flowers are borne in loose terminal heads and, like the leaves, emit the sweet scent of newly mown hay. The fruits are rough with hooked hairs. In bloom April–June.

Habitat: Widespread throughout the British Isles in deciduous mountainous woodlands, mostly in a chalk or limestone soil, especially the Chilterns, E Yorkshire, Westmorland.

History: The whole plant is rich in coumarin, present also in the Sweet Vernal Grass of meadows, which releases its fragrant herby scent as it dries after cutting. The foliage of Woodruff is almost scentless when growing, but becomes scented as it ages and, if carefully dried, will retain its fragrance for several years.

It takes its country name from the fact that it inhabits deciduous woodlands and from the Anglo-Saxon word *rofe*, denoting that its leaves formed a wheel or *ruff*. The plant was highly esteemed from earliest times, and was hung up to dry in houses as it would remain fragrant throughout the year and would help to keep the rooms cool in summer. Tusser knew it as *Sweet Grass* and recommended it for strewing, while he also advised using it in making a 'sweet water' for bathing the face to improve the complexion. Gerard said that made up into garlands 'they do make fresh the place, to the delight and comfort of such as there are there-in'. For this reason, Woodruff was used to decorate churches. In the churchwarden's accounts for St Mary-at-Hill in the City of London, there is an item for the use of Woodruff on St Barnabas' Day. Johnston said, 'the dried leaves are put amongst linen for their sweet smell', and it also acted as a deterrent to moths. It was the custom to place the sweet-smelling leaves between the pages of books, and during Georgian times the leaves were placed in the cases of pocket-watches, so that their fragrance could be inhaled whenever telling the time. The dried leaves were a welcome addition to most pot-pourris, for the longer the leaves are kept, the more powerful their scent becomes. For the same reason, they were used in snuffs and to stuff mattresses and pillows, releasing their delicious scent during the hours of sleep. 'The flowers are of a very sweet smell as is the rest of the herb', wrote Gerard; and from the blossoms and leaves a stimulating 'tea' was made which toned up the system and purified the blood. The Poet Laureate, Arthur Austin, wrote:

> She – Fresh Woodruff soaks,
> To brew cool drink and keep away the moth.

The early blooming of its flowers signified that the winter was passed, as we find in an early fourteenth-century poem called 'Springtime':

> The threstlecock him threteth oo,
> Away is huere wynter wo
> When woodruff springeth.

CROSSWORT *G. cruciata*

A hairy perennial of prostrate habit with 4-angled stems 9–12 in. tall. The pale green downy leaves are elliptic and are borne 4 in a whorl, while the pale yellow flowers appear 6–8 in axillary cymes. The upper flowers have only pistils, the lower only stamens. They are nectar-secreting and have a sweet honey-like scent. In bloom April–June.

Habitat: Widespread throughout the British Isles at the edges of woodlands, in hedgerows and by the wayside.

LADY'S BEDSTRAW
G. verum

A glabrous perennial increasing by stolons and with 4-angled stems 12–18 in. tall. The leaves, borne 8–12 in a whorl, are dark green and when dry emit the same sweet coumarin scent as Woodruff. The flowers, too, are fruit-scented and are bright greenish-yellow, borne in axillary and terminal cymes. They are visited by Coleoptera. In bloom July–September.

Habitat: Widespread on dry banks and in hedgerows to a height of about 2000 ft.

Lady's Bedstraw

HEATH BEDSTRAW
G. saxatile

A glabrous, prostrate, much-branched perennial with 4-angled stems 6–8 in. tall and obovate leaves borne 6 in a whorl. The leaves have prickles at the edges which point forward. The white flowers are borne in compact cymes and, due to the presence of indol, emit the sweet sickly scent of certain Lilies. In bloom June–August.

Habitat: Common in heathlands and moorlands and in coniferous woodlands to a height of 4000 ft; especially widespread in N England and in Scotland.

FEN BEDSTRAW   *G. uliginosum*

A glabrous perennial with a creeping rootstock and with slender upright 4-angled stems growing about 10 in. tall. The stems have prickles at the angles. The lanceolate leaves are borne 6–8 in a whorl and when dry are sweetly coumarin-scented. The flowers are white with yellow anthers and are borne in spreading panicles. In bloom July–August.

Habitat: Common in marshlands and fens throughout the British Isles.

HONEYSUCKLE FAMILY   Caprifoliaceae

Mostly shrubs with opposite, simple, exstipulate leaves. Flowers hermaphrodite, borne in a cymose inflorescence; sepals 3–5; petals 3–5, united. Corolla-lobes imbricate. Stamens 4–10, equal in number to petals and alternate with them. Ovary 3–5-chambered. Fruit, a berry or achene; seeds with endosperm. One of the most interesting of all plant families, several genera being represented in E Asia. Almost all bear sweet scented flowers and are visited mostly by Lepidoptera and to a lesser extent by Hymenoptera. *Sambucus* (Elder) pollinated by flies.

SAMBUCUS   Small trees or shrubs with stems containing pith. Leaves pinnate, stipules usually present. Flowers 5-merous, borne in panicles. Sepals 3–5; corolla with short tube. Stamens 5. Ovary 3–5-chambered with single ovule in each chamber. Style short with 3–5 stigmas. Fruit, a drupe. Pollination is by Diptera, which are attracted by the large flat surface of the tiny flowers and their short tubes. The flowers emit a strong musky smell with fishy undertones. This is more pronounced after fertilisation, and in *S. ebulus*. It is due to the presence of trimethylamine and propylamine, which give the unpleasant smell to flowers of the family Rosaceae.

The bark of *S. nigra* and also the leaves of both species contain valeric acid. This is also present in the closely related Valerian, and in perspiration, and leaves an unpleasant smell on the hands after touching.

DWARF ELDER   *Sambucus ebulus*

A shrub growing 3–4 ft tall with simple stems and lanceolate serrate leaflets. The flowers are white, tinted pink and are borne in a flat 3-branched cyme with purple anthers to attract the pollinating flies. The fruit is a black drupe. In bloom July–August.

Habitat: Rare, although found in hedgerows and on wasteland throughout the British Isles. Mostly confined to S England.

History: This is perhaps Shakespeare's 'stinking' Elder referred to in *Cymbeline,* when Arviragus says:

And let the stinking elder, grief, untwine
His perishing root with the increasing vine.

It may be a native species or may have been introduced by the Danes, for it was also known as *Danewort* or *Dane's Elder*. The plant dies back in winter.

BLACK ELDER *S. nigra* Pl. 20

A deciduous tree attaining a height of 20 ft with hollow stems and leaves divided into 4 pairs of lanceolate leaflets. The creamy-white blooms are borne in flat-topped cymes and when newly open and warmed by the sun emit a wholesome, musky 'summery' smell, resembling that of *Buddleia* flowers, without the fishy undertones of the Dwarf or Stinking Elder. The pale yellow anthers and the more pleasing smell of the flowers denote a wider range of pollinators. In bloom July.

Habitat: Widespread in woodlands and hedgerows, on waste ground near cottages and farm buildings.

History: In *Love's Labour's Lost*, Holofernes reminds Berowne of his reputation when he says 'Begin, sir; you are my elder, remember Judas was hanged on an Elder'. Sir John Mandeville, who wrote a detailed account of his travels abroad in 1356, stated that he had been shown 'the tree of elder, that Judas heng himself upon' and his writings would almost certainly have been known to Shakespeare. For this reason and also for its unpleasant smell, no plant had a more evil reputation during ancient times. Yet at the same time, the tree was planted near every cottage home as it was thought to give protection against witches. Hence, it is to be found in cottage gardens to this day.

With its hollow stems it was used to make musical instruments, including the *sambuk* (sackbut), hence its botanical name *Sambucus*, while the name *Elder* is derived from the Saxon *eller*, to kindle, for it is possible to blow through the stems to brighten a dying fire. The leaves worn in the hat will prevent flies from settling in warm weather.

In medieval times, the musk-scented flowers were placed in ale to impart their flavour, while the French put layers around early apples to give them a muscatel perfume. The fresh flowers were also dried for the same purpose, for they retain their sweet musky scent for several months. If they are distilled, the flowers yield an oil which has the scent of muscatel raisins and is present in elderberry wine, very similar to Muscat de Frontignac. 'A cup of mulled elder wine, with nutmeg and sippets of toast, just before going to bed on a cold wintry night, is a thing to be run for', wrote Cobbett, and Evelyn, in his *Sylva* (1729) said, 'It (elder wine) greatly assists longevity'.

In N Europe the Elder is more popular than it is in Britain and is named after Hulda, the goddess of love, to whom it was held sacred.

VIBURNUM Small trees or shrubs, deciduous or evergreen, growing 8–12 ft tall. Leaves simple, oval, often devoid of protective scales when in bud. Flowers white or pink, borne in cymes; corolla rotate or bell-shaped; ovary single-celled; stigmas 3. Fruit, a drupe. The flowers of a number of species have their honey readily exposed and are mostly visited by Diptera attracted by the sweet, sickly smell which is due to the presence of indol.

WAYFARING TREE *Viburnum lantana* Pl. 21
A pubescent deciduous shrub growing 12 ft tall with dark green heart-shaped leaves, serrate at the edges and downy on the underside. The creamy-white flowers are borne in a flat-topped umbel and have the scent of Lilies. In bloom May.
Habitat: By the wayside and in woodlands, in calcareous soils; most common in SE England and E Yorkshire.
History: It was so named by countrymen as it provided dense shade in the hedgerow for anyone wanting to rest for a while under its branches and inhale its sweet perfume.

GUELDER ROSE *V. opulus* Pl. 21
A small tree growing about 10 ft tall with grey-green twigs and acrid bark. The leaves are 3–5-lobed with glandulous hairs on the stalk and in autumn take on rich shades of crimson and bronze. The white flowers are borne in flat umbels, the outer flowers sterile with petal-like corollas. They have a slight mossy perfume and are followed by brilliant red fruits. In bloom June–August.
Habitat: Unlike *V. lantana* it prefers a moist situation and is present in shady woodlands and in hedgerows, especially in the fenlands of England and Wales.

LINNAEA Slender, creeping evergreen. Leaves small, exstipulate. Flowers borne in pairs on long stems. Corolla funnel-shaped; petals 5, unequal. Stamens 4, 2 longer than others. Ovary with 1 fertile and 2 sterile cells. Stigma capitate. Fruit, an achene.

TWINFLOWER *Linnaea borealis*
A prostrate evergreen with stems 5–6 in. tall and covered in downy hairs. The small, ovate leaves are thick and leathery, while the drooping flowers are borne in pairs on long peduncles. The pendulous position protects the honey from rain, and the hairs on the interior of the corolla exclude small flies. The five purple lines inside the corolla direct butter-flies and the larger Diptera to the honey. The flowers are deep pink, crimson on the inside, and diffuse a soft honey-like scent when the sun shines on them. In bloom July–August.

Habitat: Very rare in a few pine woods in Northumberland and in N Scotland, almost in sight of *Aurora borealis* (Northern Lights).
History: It is 'The little northern plant, long overlooked, depressed, abject' which the Swedish botanist, Carl von Linné (Linnaeus 1707–1778), described and named after himself.

LONICERA   Erect shrubs, sometimes twining, with leaves entire, exstipulate. Flowers borne in axillary pairs or in cymes, often united by their ovaries. Corolla tubular and 2-lipped with 4-lobed upper lip and entire lower lip. Stamens 5. Ovary 2-celled. Style slender; stigma capitate. Fruit, a berry. *L. caprifolium* pollinated by humming-bird hawkmoths at night which are the only insects able to reach the honey secreted at the base of the 30-mm tubes. The native *L. periclymenum* has a 22-mm tube and the honey, which rises in this tube, is therefore accessible also to the longest-tongued bee, *Bombus hortorum*, which has a 2-mm proboscis and will effect fertilisation if there are few moths about. Both species bear creamy-white flowers which are highly scented by day and at night.

HONEYSUCKLE, WOODBINE   *Lonicera periclymenum*
Pl. 20
An erect perennial with twining stems and ovate leaves, glabrous or pubescent and untoothed. They appear early in the year. The richly scented flowers are borne in terminal heads and are creamy-yellow, shaded with red on the outside, usually becoming darker after pollination. They are followed by a globular crimson berry. In bloom June–September.
Habitat: Common in woodlands and hedgerows throughout the British Isles.
History: It was named *Lonicera* in honour of Adam Lonicer, a physician of Frankfurt. William Turner in his *Names of Herbs* (1548) wrote, 'periclymenon is named in English Woodbynde and in some places of England, honeysuccles'. As Gerard said, 'the flowers show themselves at the top of the branches, many in number, long, white, sweet of smell . . . with threads growing out of the middle'. The plants turn themselves towards the sun, as Shakespeare observed, and it is only when 'ripen'd by the sun' that the flowers emit the fullness of their aromatic clove-like perfume which at night penetrates for some distance.

William Bullein in his *Book of Simples*, published during the reign of Henry VIII, wrote, 'how friendly does this herb if I may so name it, embrace the bodies, arms and branches of trees with its long winding stalks and tender leaves or spreading forth his sweet lilies'. In *A Midsummer Night's Dream* Shakespeare speaks of the plant in similar endearing terms when Titania tries to bestow sleep upon her companion Bottom:

> Sleep thou, and I will wind thee in my arms,
> . . . so doth the woodbine, the sweet honeysuckle
> Gently entwist; the female ivy so
> Enrings the barky fingers of the elm.
> O, how I love thee! How I dote on thee!

Throughout his plays, Shakespeare makes it clear that the Woodbine is the Honeysuckle and not *Clematis vitalba*, erroneously called the *Woodbine* by many writers and poets. *L. periclymenum* has always been known as the emblem of love and affection because of the marks it makes on other plants as it twines about them. The Scottish poet James Graham said:

> . . . round the brier, the honeysuckle wreaths
> Entwine, and with their sweet perfume, embalm
> The dying rose.

Tennyson speaks of its scent at night-time in some of the loveliest lines ever written:

> Good Lord, how sweetly smells the honeysuckle
> In the hush'd night, as if the world were one
> Of utter peace, and love, and gentleness.

Thomas Moore in his poem 'Evening Perfume' also mentions its scent at night:

> 'Twas midnight – through the lattice, wreath'd
> With woodbine, many a perfume breath'd
> From plants that wake when others sleep . . .

Ben Jonson wrote of its use in making scented arbours, where one could rest for a while and which were to be found in most cottage gardens of the time:

> . . . the blue bindweed doth itself enfold
> With honeysuckle and both these entwine
> Themselves with briony and jassamine,
> To cast a kind and odoriferous shade.

PERFOLIATE HONEYSUCKLE *L. caprifolium*

A glaucous twining perennial distinguished from *L. periclymenum* in having the upper leaves united at their bases. The ovate leaves are dark green, glaucous on the underside. The flowers are borne in whorled heads and have an extended corolla-tube so that they are fertilised entirely by night-flying Lepidoptera. They are in bloom when night hawkmoths are most abundant. The flowers are creamy-white, paler than those of *L. periclymenum* and are followed by globular red berries. In bloom May–June.

Habitat: Native to S Europe, it reached Britain early in her history and is extensively planted in gardens as it comes into bloom early and thus extends the Honeysuckle season. It is naturalised in hedgerows in S and E England but is rare elsewhere.

MOSCHATEL FAMILY   Adoxaceae

Glabrous perennial herb with creeping rhizomatous rootstock. Root leaves 1–3 and 3-lobed. Cauline leaves opposite, in pairs. Greenish-white flowers borne in a long-stalked terminal head of 5. Terminal flower with 2-lobed calyx; corolla 4-lobed; stamens 4, alternating with corolla-lobes. Lateral flowers with 3-lobed calyx; 5-lobed corolla; stamens 5. Ovary 3–5-celled with single ovule in each cell. Fruit, a drupe.

MOSCHATEL   *Adoxa moschatellina*

A perennial, with fleshy pointed scales at the top of the rhizomatous rootstock. The stem is 4-angled and grows 4 in. tall, while the trifoliate leaves are borne on 3-in. stems. Each flower-stem has 2 cauline leaves or bracts and 1 in. above appears the flower head, arranged in a cube but with a tiny terminal flower above. Upon approaching, the whole plant emits a muscat scent, to some people resembling almonds, to others elder blossom. Pollination is by small flies which are attracted to the smell and suck the nectar secreted at the base of the stamens. In bloom April–May.

Habitat: Widespread throughout England, Scotland and Wales (rare in, or absent from, Ireland) in woodlands and hedgerows and on mountainous pastures to a height of nearly 4000 ft.

VALERIAN FAMILY   Valerianaceae

Herbs, sometimes woody, with strong-smelling rhizomatous roots. Leaves opposite or radical, exstipulate. Flowers hermaphrodite or unisexual, borne in cymose panicles. Calyx superior, variously toothed; corolla of 3–6 united petals, tubular or spurred at base. Stamens 1–4, inserted at base of tube. Ovary 1–3-chambered; fruit dry, indehiscent. The roots aromatic or unpleasant-smelling. *Valeriana jatamansi* of Bhotan, all parts of which are fragrant, furnished the spikenard of the Scriptures which Mary used to anoint the feet of Jesus. As it grew in only one region of the world, about 17,000 ft above sea-level, it was so precious that a small alabaster box of spikenard ointment was valued at about £10 in today's money value. It was used by the Romans to anoint important guests during feasts. Horace promised Virgil a cadus (about 50 bottles) of wine in exchange for a small box of this precious ointment.

The woody rootstock is covered in hairy fibres caused by the petioles of withered leaves, and it is from the portion immediately above the root that the best spikenard is obtained. It comes from the Himalayas in bundles, with each piece of finger thickness, grey in colour and surrounded by fibres, like the tail of a sable. The odour increases with age, as the root dries, and becomes 'heavy', like Patchouli. It is the Mountain or Indian Nard of Dioscorides and was the most valued perfume of the ancient world.

VALERIANA Perennial herbs, glabrous or slightly pubescent. Leaves entire or pinnatifid. Flowers bracteolate, borne in capitate cymes, red, pink or white in colour. Corolla funnel-shaped, pouched at the base. Stamens 3; fruit a single-chambered nut.

Valeric acid, also found in perspiration, is present in the British species and gives all parts of the plant, particularly the roots, its characteristic rancid smell.

COMMON VALERIAN *Valeriana officinalis*

A glabrous perennial growing 2–4 ft tall with dark green pinnate leaves and lanceolate leaflets, often toothed. The pale pink flowers are borne in a terminal cyme, with the tube of each floret 5 mm long, and are visited mostly by daytime Lepidoptera. In bloom July–August.

Habitat: Common in moist woodlands, hedgerows and meadows; also in dry habitats on chalk and limestone.

History: It takes its name from the Latin *valeo*, to be well, because of its valuable medicinal qualities. Its country name *All-Heal* and its dedication to the Blessed Virgin originated from these qualities, despite the nauseous rancid smell of stale perspiration given off by its roots. Cats are fascinated by the roots and become almost intoxicated if they nibble at them. The roots were taken for nervous exhaustion brought on by emotional excitement, while the essential oil lessens the sensibility of the spinal cord after nervous attacks.

When exposed to air, the essential oil becomes oxidised to form valeric acid which is the source of the abominable smell. The root also contains the unpleasant-smelling acetic and formic acids.

TEASEL FAMILY Dipsacaceae

Herbaceous plants with opposite exstipulate leaves. It is closely related to Compositae which the flowers resemble in their aggregation and the accessibility of their honey. The florets are crowded together with an involucre of bracts. Calyx superior, expanding into a cup-shaped tube; corolla tubular with 4–5 unequal lobes. Stamens 4, anthers free. Fruit dry, indehiscent. Few have any scented attractions; this is unnecessary as nectar is easily accessible.

KNAUTIA Annual or perennial herbs with leaves opposite, mostly pinnatifid though some undivided. Flowers borne in convex heads with an involucre of lanceolate bracts. The flowers increase in size from the centre due to development of the outer lobe of the corolla. In central florets the tube is about 4 mm long with corolla-lobes similar, while in the marginal florets the tube may be up to 9 mm long. Thus, almost all types of insects visit the flowers because of their continuous and gradual development. The honey secreted by the ovary is protected from rain by a ring of hairs.

FIELD SCABIOUS *Knautia arvensis*
Syn: *Scabiosa arvensis*
A branched perennial herb growing 2–3 ft tall with pinnately lobed lower leaves. The flowers, just over an inch wide, resemble pin-cushions with their tightly packed florets and are borne on wiry stems about 2 ft in length. They are a lovely shade of soft lilac-pink with a delicate sweet honey scent, and are visited mainly by butterflies. In bloom July onwards.
Habitat: Widespread by the roadside, in hedgerows and cornfields, especially in calcareous soils and mostly in the Midlands and E England.

DAISY FAMILY Compositae

Mostly herbs (rarely trees or shrubs) accounting for 10% of the earth's species of flora and distributed over a larger part of the earth's surface than any other plants. Leaves usually alternate (opposite in the Eupatorieae family), exstipulate, occasionally whorled. Root a tap-root or tuber as in *Dahlia*, thick as in *Daucus* (Carrot). Flowers arranged in heads, flat or convex, divided into racemes or corymbs and surrounded by scales or bracts forming an involucre. Florets may be similar as in *Carduus* (Thistle), or outer florets may differ from inner ray-florets as in *Bellis* (Daisy). Calyx of 5 sepals, limb represented by pappus of hairs. Calyx variable but usually tubular or ligulate (extended on one side) with 5 equal teeth. Florets perfect with 5 stamens as in Dandelion; or stamens absent as in Daisy. Carpels 2, united to form a 1-chambered inferior ovary with single style. Anthers cohere to form a hollow tube and dehisce introrsely, filling the tube with pollen before the flower opens. The stigmas lie close together in the base of the tube and as the style grows brush the pollen from the tube by means of minute hairs on their surface, thus exposing the pollen to insect visitors.

Because of their aggregation the flowers are conspicuous to insects, while the easy accessibility of honey ensures that they are visited by all sorts of pollinators. In Umbelliferae, the honey lies on the flat disc exposed to the rain and is visited only by the shortest-tongued insects, thus the flowers mainly have an unpleasant smell. In Compositae, the honey is secreted by a ring surrounding the style at the base of the corolla-tube and as it accumulates it rises in the tube where the anthers shelter it from rain. The honey is therefore available to all forms of insect life, and in many species the flowers have an attractive musky scent as in *Calendula*, *Carduus nutans*, or a sweet almond-like perfume as in Purple Hawkweed and *Petasites fragrans*. The leaves of several species have an aromatic perfume, while the roots of others become scented as they dry, e.g. Elecampane.

SENECIO Annual or perennial herbs or shrubs, some climbing, often with fleshy stems. Leaves, deeply toothed, spirally arranged. Flowers yellow, borne solitary or in corymbose heads with bracts arranged in a single row. Ray-florets usually absent, disc-florets tubular. Pappus of several rows of soft, silky hairs.

STINKING GROUNDSEL *Senecio viscosus*
An annual with viscid stems covered in glandular hairs and with dark green deeply pinnatifid leaves which are also viscid. The long-stalked flowers are borne in a compound corymb with the outer bracts almost half as long again as the inner; the ray-florets short. The whole plant has an unpleasant fetid smell and is visited mainly by dung-flies and midges. In bloom July–September.
Habitat: Widespread on waste ground near the coast and on railway sidings and embankments. Rare in Scotland and in Ireland.

PETASITES Perennial herbs with a rhizomatous rootstock and large radical leaves. They usually appear after the flowers which are borne in a spike-like raceme. Florets tubular; female spikes with female marginal florets and 1–5 central sterile florets. Pappus present in females, rare in sterile florets. Achenes cylindrical.

WHITE BUTTERBUR *Petasites albus*
Syn: *Tussilago alba*
A downy perennial with a rhizomatous rootstock and leaves almost 12 in. across, cordate, with large marginal teeth and held on long stalks. The flowers are borne on stems 2 ft tall and are pure white with pale green bracts. They are sweetly scented, especially at night. Pollination is by butterflies and moths. In bloom April–May.
Habitat: Naturalised on waste ground and at the margin of woodlands especially in the Midlands, but also as far north as Aberdeen.

WINTER HELIOTROPE   *P. fragrans*   Pl. 22
Syn: *Tussilago fragrans*
A creeping perennial with large leaves like those of the closely related
Coltsfoot, with long stalks and pubescent on the underside. They are
retained through winter until after the flowers have appeared, when new
leaves take their place. The flower-spikes are composed of 10–12 florets
of palest lilac-mauve and have a sweet almond-like scent, like Heliotrope. They are visited by bees in the absence of butterflies which are not
about in the British Isles when the plant is in bloom December–March.
Habitat: A garden escape to wastelands and often present on the banks
of streams and rivers, usually in S England.
History: Though now naturalised in several parts of England and in
Ireland, the plant was unknown outside its native Italy until discovered
in 1800 by M. Villan, growing at the foot of Mt Pilat and scenting the
warm air around. It was grown in pots in Paris to perfume the winter
gardens of the aristocracy and reached England for a similar purpose
shortly afterwards. It came to be widely planted in gardens, though
with its creeping habit it soon became an obnoxious pest, covering the
ground with its large leaves to the exclusion of all other plants. In pots,
however, it may be kept in bounds and used to perfume the home. Out
of doors it should be planted on waste ground or beneath trees where
little else will grow. It takes its name from the Greek *petasos*, a broad-
brimmed hat, from the large size of its leaves.

INULA   Erect annual or perennial plants with leaves spirally
arranged and flowers borne solitary or in corymbs. Bracts imbricate,
arranged in several rows; ray-florets in single row; disc-florets tubular.
Anthers with 2 bristles at base. Pappus of single row of hairs.

ELECAMPANE   *Inula helenium*
A stout perennial, growing 4–5 ft tall with a tuberous rootstock. The
leaves are 12 in. long, deeply wrinkled and downy, toothed at the
margins. The upper leaves clasp the stem, while the lower are long-
stalked. The handsome flowers, some 3 in. across, are borne solitary or
in twos and threes and are brilliant golden-yellow with downy ovate
bracts and narrow spreading ray-florets. Gerard said that 'the stalk is
about a finger thick, divided at the top into divers branches; at the top
of every sprig stand great flowers, broad and round of which not only
the long small leaves which compass round about are yellow, but also
the middle ball or circle which is filled with an infinite number of
threads'. He added, 'the root is thick, as much as a man may grip, not
long, blackish without, white within, full of substance, sweet of smell,
and bitter of taste'. Only the root is scented, with a pronounced banana

smell when freshly dug, which takes on the violet scent as it dries, later becoming aromatic. In bloom July–August.

Habitat: It is found throughout the British Isles but is comparatively rare and is most common in S England. Present in meadows and copses and by the wayside.

History: It may be a native plant or may have been introduced, but by Gerard's time it was well established for he gives various places where it was known to him, though he does not speak of it as a garden plant. 'It groweth plentifully', he said, 'in the fields as you go from Dunstable to Puddlehill; also in an orchard as you go from Colebrook to Ditton Ferry, which is the way to Windsor'. William Coles said that it 'is one of the plants whereof England may boast as much as any for there grows none better in the world than in England'.

It takes its botanical name from the Greek *elenium* which is derived from *helenium*, a plant commemorating Helen of Troy. Thus both names have the same meaning. Dioscorides described its virtues and Pliny said that 'Julia Augusta let no day pass without eating some of the roots of Ennula, candied'. In the Middle Ages it was candied and eaten as a medical sweetmeat to help the digestion. Indeed, until the end of the nineteenth century, flat cakes of Elecampane (*Enula campana*) could be obtained in most apothecaries' shops in London.

The root contains a crystalline principle resembling camphor, called helenin, also a starch inulin, a volatile oil which has the scent of labdanum and a trace of acetic acid which may be detected immediately the root is lifted and broken. Helenin is a powerful antiseptic and was said to kill the bacillus of tuberculosis. It was also used to ease bronchitis. In France the root is used in the distillation of absinthe.

PLOUGHMAN'S SPIKENARD *I. conyza* Pl. 22

Syn: *Conyza squarrosa*

An erect downy perennial growing 3–4 ft tall with stems branched at the top. The lanceolate leaves, about 6 in. long, are downy and toothed at the margins. The flowers are borne in a branched corymb and are small and dingy yellow, with narrow outer bracts recurving at the tips; the inner bracts are of purple colouring. In bloom July–September.

Habitat: Widespread on calcareous soils, especially in the Midlands, Northumberland and S Wales. Present in scrublands and on cliffs and rocky ground.

History: Though not as well-known as Elecampane, its root has an aromatic spicy scent and was uprooted, dried and hung about the cottager's home to scent the musty air. In his poem dedicated to Cowper Green, John Clare describes his joy at finding the plant there (p. 17).

ANTHEMIS Annual or perennial herbs with hairy stems and pinnate leaves with an aromatic scent. The leaves, spirally arranged, are 1–3-times pinnatifid; the flower-heads solitary with imbricate bracts in few rows; ray-florets in single row, mostly white; disc-florets tubular. Pappus a membranous ring.

CORN CHAMOMILE *Anthemis arvensis*

A much-branched hoary annual growing about 12 in. tall with leaves divided into oblong hairy segments which give the plant a greyish appearance. The leaves emit a slight fruity scent when handled but the flowers have a more pronounced aromatic fragrance. The ray-florets are white with yellow disc-florets and are borne in long-stalked conical heads. Pointed lanceolate bracts are to be found between the disc-florets. In bloom May–August.

Habitat: Common throughout the British Isles, found mostly in sandy soils and at the edges of cornfields, also on waste ground.

COMMON CHAMOMILE *A. nobilis*

A prostrate perennial with a much-branched stem and forming a dense mat when established. The leaves are almost free from down and are cut into hairy segments. When bruised or trodden upon they emit a powerful aromatic scent. The flowers are borne solitary and are long-stalked, the bracts downy with white margins; the ray-florets white and spreading, the disc-florets yellow. The flowers droop before they open. In bloom June–August.

Habitat: Common in pastures, by the wayside and on waste ground of a sandy nature throughout the British Isles, though less common in Scotland.

History: It was used in earlier times to make a fragrant 'lawn', for as Falstaff remarked to Henry, Prince of Wales, soon to become King Henry V: 'Harry, I do not only marvel where thou spendest thy time, but also how thou art accompanied; for though the camomile, the more it is trodden the faster it grows, yet youth, the more it is wasted the sooner it wears.'

Lawson in *The New Orchard* wrote, 'large walks . . . raised with gravel and sand, having seats and banks of camomile; all this delights the mind and brings health to the body'. In the early seventeenth-century play *The More the Merrier* we find:

> The camomile shall teach thee patience
> Which riseth best when trodden most upon.

Parkinson compared the pungency of its deeply serrated leaves with that of the Featherfew, though from its scent it derives its name from the Greek for earth apples. In Spain it is known as *Manzinella*, little apple. During Elizabethan times, before the introduction of tobacco, its leaves were dried and smoked, and the rich aroma was a cure for sleeplessness. Its flowers are used for making a drink: one ounce of the blooms infused in a pint of boiling water and a wine-glassful taken twice a day soothes tired nerves and relieves indigestion. The same preparation may also be used as a hair tonic, while the celebrated Dr Schimmelbusch recommended it as a mouthwash.

When making an infusion, the daisy-like flowers, which are also scented, should be removed when just fully open, as it is the centre of these which contains the medicinal qualities.

Countrymen would infuse the leaves in spirit of wine to rub on their arms and face when working in the fields to keep away flies and wasps. Oil of chamomile is distilled from all parts of the plant, the flowers yielding rather less than 1% of oil.

To make a bath to revive the spirits, 'Take mallows, pellitory-of-the-wall, of each, 3 handfulls; camomell flowers, mellilot flowers, of each 1 handful; of hollyhock flowers, 2 handfulls; Isop, one great handful; senerick seed, one ounce. Boil in 9 gallons of water, add a quart of new milk and to into it blood warm'.

To make a 'lawn', plants may be raised from seed sown in early summer, with a small packet of seed producing almost a thousand plants. Sow in drills and keep the soil moist. If the seed is sown thinly, transplanting will not be necessary until the 'lawn' is about to be made in autumn. Where the creeping Thymes and possibly the creeping Mints are also to be planted, set the Chamomile 12 in. apart and as they are fairly inexpensive and hardy use a greater number of these than of other plants.

After making the lawn the plants may be rolled in areas where the soil is light and well drained. John Evelyn said that 'it will now (October) be good to beat, roll and mow carpet walks of chamomile'. The ground must be kept weeded during the first year, until the plants spread out and prevent annual weeds from forming. As the Chamomile is evergreen, the lawn will not lose its green colour during winter, but expect it to have rather a miserable appearance until the plants have become established. Until then one should not walk over the plants except when weeding. Once established, they will remain green during the hottest weather and will tolerate any amount of treading.

Towards the end of summer the plants may be clipped back hard, though not so late as to prevent them becoming green again before winter. A light clipping, say twice a year, will suit them best and it is a good idea to roll them whenever the ground conditions are suitable. Once established, only occasional weeds will need removing, and the lawn should be extremely long lasting, for only the young newly set plants show any tendency to die. For this reason a few plants should be held in reserve for filling any gaps, or pieces may be removed from established plants instead.

Besides the value of its medicinal properties, the Chamomile frequently figures in folklore. In Germany it is thought that if a wreath of Chamomile is hung up in the home on St John's Day it will guard against thunder and lightning. William Browne, writing in the *Britannia's Pastorals*, expressed his belief that Chamomile was good for fish:

> Another from her banks in sheer good will,
> Brings nutriment for fish, the camomile.

Because of its refreshing fruity scent it was used in ancient times to make chaplets and garlands:

> Diana!
> Have I (to make thee crowns) been gathering still,
> Fair-cheek'd eteria's yellow camomile?

Because of its prostrate form, the plant was the symbol of humility to the early writers.

STINKING CHAMOMILE   *A. cotula*

An erect annual growing 2–3 ft tall and with branched stems, the leaves glandular, dotted and divided into hairy segments. They emit an unpleasant, sickly, rancid smell when bruised and will cause the hands to blister if handled excessively. The flowers have white ray-florets which turn down after fertilisation, and yellow disc-florets.
Habitat: Common in cornfields and wasteland, especially in S and C England; rare in Scotland.

MATRICARIA Annual herbs with much-branched erect stems and leaves divided into narrow linear segments. Flowers usually lacking ray-florets; the bracts imbricate, arranged in several rows. Pappus absent.

WILD CHAMOMILE, SCENTED MAYWEED
Matricaria chamomilla
Syn: *M. recutita*

A much-branched glabrous annual, the pinnate leaves divided into hairy segments which end in bristle-like points. They are pleasantly aromatic when handled and release their scent without bruising. The flower-heads are conical, the bracts without membranous margins, with the ray-florets turning down or possibly absent altogether. Distinguished from *M. inodora* by its scent and its bracts. In bloom June–September.

Habitat: Common on arable land and by the wayside throughout England and Wales.

History: It takes its name from the Latin *mater cara*, beloved mother, for it is dedicated to St Anne, mother of the Blessed Virgin. Its medicinal qualities are as valuable as those of *Anthemis nobilis*. Commercial oil of chamomile is derived mainly from this plant, which is grown in S Europe on a considerable scale for this purpose. The oil contains caprinic acid which is colourless and has the fruity chamomile scent.

PINEAPPLE WEED, RAYLESS MAYWEED

M. matricarioides
Syn: *M. suaveolens*

A hairless annual growing 9–15 in. tall, much-branched above and with dark green leaves divided into thread-like segments. The flowers are borne solitary on short stalks; the bracts are blunt with white margins, ray-petals are absent. The whole plant has the rich, refreshing fruity smell of pineapples. In bloom June–September.

Habitat: Possibly introduced at an early date, it is common on paths and waste ground and at the edges of corn-fields throughout the British Isles, usually growing in colonies in sandy soil, and scenting the air around.

Pineapple Weed

ACHILLEA Erect perennials with leaves spirally arranged and usually 2-pinnatifid. The flowers are borne in corymbs, the bracts imbricate, arranged in several rows, the ray-florets white or yellow and broad; the disc-florets also white or yellow, tubular. Pappus absent.

YARROW, MILFOIL *Achillea millefolium* Pl. 23

An aromatic perennial with erect downy stems 12–15 in. tall and slightly hairy bi-pinnate leaves, the dark green leaflets cut into hair-like segments. The flower-heads, white or pink, are borne in flat terminal umbels. Each flower has 5 broad triangular ray-florets, with white disc-florets and black styles. It is visited by all types of insects which seek the honey secreted at the base of the style. In bloom June–August.

Habitat: Common in meadows and on grassy banks throughout the British Isles.

History: Its name *Yarrow* is a corruption of the Greek *hiera*, holy herb, for it possessed numerous medicinal qualities. Its botanical name is that of the Greek warrior *Achilles*, who is said to have cured the wounds of his soldiers with the leaves of the plant, which have since been frequently used by countrymen for this purpose.

So efficient were the leaves that the plant was called *Woundwort* and *Carpenter's Weed*, for a decoction of the leaves was used to stem the flow of blood caused by carpenters' tools.

Both the flowers and leaves have an aromatic scent and yield a dark green volatile oil; also achilleic acid.

If the plant is macerated in water and allowed to ferment, a blue oil floats to the surface, leaving behind a residual water which, with the addition of salt, is agitated with ether, and leaves a brown oil of aromatic odour. Eucalyptol is also present to give the plant its familiar 'herby' scent.

CHRYSANTHEMUM Annual or perennial herbs or shrubs of erect habit. Leaves toothed or lobed, arranged spirally. Flowers borne solitary or in corymbs, bracts imbricate with membranous margins. Ray-florets white or yellow, arranged in a single row or absent; disc-florets tubular. Pappus absent. Achenes cylindrical.

OX-EYE DAISY, MOON DAISY, MARGUERITE
Chrysanthemum leucanthemum
An erect woody perennial herb, almost glabrous and growing 1–2 ft tall. The leaves are dark green, the upper leaves lanceolate, the lower long-stalked and lobed. The 2-in. flowers are borne solitary, the bracts having a purple membranous margin. The 20 ray-florets are tubular and brilliant white, each having an abortive stamen fitted with a white lobe. The disc is made up of as many as 500 yellow florets into which the honey rises to a depth of 1 mm so that it is accessible to the shortest-tongued insects, the flies, which are attracted to the flowers by their slight unpleasant smell of perspiration. In the absence of insects, self-fertilisation is possible. In bloom June–August.

Habitat: Widespread throughout the British Isles and N Europe, in meadows and on wasteland.

History: The plant was named *Maudeline* from *St Mary Magdalene* and was also called *Moon Daisy* from its likeness to a full moon. The young leaves were used to flavour soups and stews, while the boiled juice, sweetened with honey, was taken to relieve a hard cough. It is believed to be the flower mentioned by Chaucer in *The Flower and the Leaf*:

1

2

Plate 21
Honeysuckle
Family (ii)   1 Guelder Rose   2 Wayfaring Tree

Plate 22
Parsley Family
Daisy Family

1 Coriander
3 Ploughman's Spikenard

2 Caraway
4 Winter Heliotrope

And at the laste there began anon
A lady for to sing right womanly
A bargaret in praising the Daisie:
For – as methought among her notés sweet,
She said, 'Si doucet est la Margarete.'

SHASTA DAISY *C. maximum*

A perennial of erect habit and forming numerous leaf-rosettes. The plants grow 2–5 ft tall with long-stalked lanceolate leaves, bluntly toothed and bearing single flowers up to 4 in. across. A number of garden varieties have been introduced with attractively twisted petals, and in 1932 Esther Read appeared, the first variety bearing double blooms. Others of merit include Horace Reed and E. T. Killin, which has greenish disc-florets. The first yellow-flowering form, named Cobham Gold, was discovered in the gardens of Cobham Hall, seat of the Earl of Darnley, in 1950. In bloom May–September.

Habitat: Occasionally on waste ground near towns and villages as garden escapes. Native to SW Europe.

FEVERFEW, FEATHERFEW   *C. parthenium*

A pubescent branched perennial growing 1–2 ft tall with pinnate leaves, the 1–2 narrow leaflets divided into lobed segments and a bright yellowish-green in colour. The lower leaves have long stalks, the upper are short-stalked. The flowers, about an inch across, are borne in terminal clusters, with white ray-florets and yellow disc-florets. The whole plant emits an aromatic pungent scent when handled which is disliked by bees. In bloom June–September.

Habitat: Common in dry places throughout the British Isles, at the sides of paths and on old walls.

History: Its name *Feverfew* is a corruption of *Febrifuge*, alluding to its well-known properties for curing agues and fevers. In the West Country, the garden variety, with its double blooms, is known as *Batchelor's Buttons*. Gerard said that if the leaves were fastened to the wrists they would prevent ague. It is also known as *Featherfew* because of the feathery appearance of the leaves which retain their bright golden-green colour during winter. Its essential oil contains camphor and acts as a tonic to the nervous system. It will also bring immediate relief if it is applied to insect bites. In Italy, the leaves are fried with bacon and eggs to remove excess grease, while the dried leaves, placed in muslin bags among clothes, will keep away moths.

TANSY   *C. vulgare*   Pl. 23

Syn: *Tanacetum vulgare, C. tanacetum*

A perennial growing 2–3 ft tall with branched angular stems. The handsome dark green fern-like leaves are divided into numerous pairs of deeply pinnatifid leaflets with serrated edges, and emit a powerful aromatic smell when handled. The essential oil is stored in minute glandular dots which release the camphor-like scent when pressed. The tiny button-like flowers are borne in flat-topped terminal heads with marginal bracts and yellow disc-florets. Ray-florets are short or absent. In bloom July–September.

Habitat: Present in hedgerows and on wasteland throughout the British Isles.

History: It takes its name *Tansy* from *Athanasia* (*A. tan-asis*), the ancient word for immortality, because it remains so long in bloom. It is dedicated to St Athanasius and was a much-loved plant as Michael Drayton said in the *Muses' Elysium*:

> Then burnet shall bear up with this
> Whose leaf I greatly fancy,
> Some chamomile doth not amiss
> With savory and some tansy.

At Eastertime, its leaves were used to add their pungency to fried cakes as an alternative to the more expensive cinnamon and nutmeg. An old Easter carol ran:

> Soon at Easter commeth Alleluya,
> With butter, cheese and tansy.

The leaves were finely shredded, beaten up with eggs and fried in flat cakes called *Tansy-cakes* or *Tansies*. They were eaten during Lent in remembrance of the bitter herbs eaten at the Passover. The leaves will also add a touch in piquancy to a salad, though they should be used sparingly. In medieval times when there was no cold storage, Tansy leaves were used to rub on meats to keep flies away.

Isaak Walton speaks of a tasty dish consisting of minnows fried in egg yolk with the flowers of Cowslips and Primroses and a sprinkling of Tansy; 'thus used, they make a dainty dish'.

Tansy-cakes were served at the Coronation feast of James II and his Queen, Mary of Modena, to accompany some fifteen hundred 'dishes of delicious viands' among which were '4 fauns; stag's tongues; gotwits; brown buds and taffeta tarts'.

All parts of the plant are aromatic and upon distillation with water will yield an essential oil with a minty smell. This contains an aldehyde which, with bisulphite of sodium, forms a crystalline compound known as tanacetone, also present in the essential oil of sage, wormwood and thuya.

The curled-leaf form, *foliis crispis*, is more attractive as a garden plant and is more highly scented than the common form.

COTULA   Glabrous annual or perennial water-plants with alternate, deeply pinnate leaves. Flowers rayless, borne in long-stalked heads.

BUTTONWEED   *Cotula coronopifolia*
A glabrous annual with sheathing leaves, deeply pinnate and with narrow lobes. It is a plant of semi-creeping habit and sends up its flowering stems to a height of 9 in. The button-like flowers are borne in small solitary heads, the disc-florets having a 4-lobed corolla-tube with the marginal florets arranged in a single row. Pappus is absent, or almost so. The whole plant has the camphor-like smell of Tansy. In bloom July–August.
Habitat: Present in marshlands and on river-banks in a few areas of England including the Wirral Peninsula and near the coast of Lancashire. It may have been introduced by seamen, for it is a common weed of the Cape Peninsula.

ARTEMISIA   Perennial, rarely annual herbs, downy or glabrous. Leaves scattered, pinnatifid. Flowers small, borne in racemes or panicles; bracts with membranous margins; disc- and marginal-florets tubular. Pappus absent. The anther-tube projects beyond the corolla with the pollen caught on the hairy stigmas of the style and exposed to the wind, a characteristic of only the most primitive plants. It is not necessary, therefore, for the flowers to be scented, though the leaves of several species are pleasantly aromatic.

MUGWORT   *Artemisia vulgaris*
An erect, slightly pubescent perennial growing 3–4 ft high with grooved and angled stems and dark green 1–2-pinnate leaves with the leaflets deeply toothed and covered with down on the underside. The flowers are yellowish-brown like those of Mimosa and are borne in panicles or spikes with downy bracts. In bloom July–September.
Habitat: Common by waysides and in hedgerows throughout the British Isles.
History: It was named *St John's Wort* for people believed that John the Baptist wore a girdle of it when he was in the wilderness. This may well be so for it grows in the East and will keep away flies if it is worn, and moths if it is placed between linen and clothes. Those who had to walk long distances would place the slightly scented leaves in their shoes to

prevent weariness, even though 'one may walk forty miles a day'. Geese and various meats were stuffed with the herb which contains the bitter principle absinthin; its essential oil is not aromatic, however. It was used to impart its bitterness to ale and, like Ale-hoof, would clarify and preserve it.

CHINESE OR VERLOT'S MUGWORT *A. verlotorum*

A perennial with a rhizomatous rootstock and a stem which is covered in pubescence. The leaves are dark green, more elongated than those of *A. vulgaris* and less downy on the underside. The flowers are borne in leafy panicles, when those of *A. vulgaris* have finished. The whole plant is more highly aromatic than the Mugwort and the leaves may be dried and used in scent-bags and pot-pourris. In bloom October–November.

Habitat: Introduced from China during the seventeenth century, it is naturalised on waste ground in SE England, especially in the London area.

WORMWOOD *A. absinthium* Pl. 23

A handsome perennial growing 12 in. tall with its erect angled stems and leaves densely covered in down. The 2-pinnatifid leaves have blunt segments or lobes and are highly aromatic. The dull yellow globular flowers are borne in elegant panicles. In bloom July–August.

Habitat: Common on waste ground and by the roadside throughout the British Isles. In parts of Texas and Mexico vast areas are covered with the plants which form an almost impenetrable jungle of interlacing stems, while the countryside takes on a greyish hue when seen from afar.

History: It retains its lovely grey-green foliage through winter and should be in every shrubbery, but it possesses an aromatic bitterness quite distinct from all other herbs. It is used in the manufacture of absinthe and Chartreuse and as a tonic beer.

All parts of the plant are extremely bitter and in 1538 Thomas Tusser wrote:

> Where chamber is swept and Wormwood is strowne,
> No flea for his life, abide to be known.

Strewn over the earthen floor of cottages and manor houses, it would keep away all insects. Tusser wrote of the qualities of the herb on numerous occasions, for the plant was highly valued by countrymen. He mentioned its use 'in places infected' and as 'a comfort for heart and the brain' (p. 20).

In medieval times the plant was, with few exceptions, esteemed above all others for its medicinal qualities, while it was one of the few plants known to counteract the poisonous effects of Hemlock and toadstools. In addition, a fermentation of the leaves could be used to remove bruises, and it was also placed among furs to keep away moths and fleas. A few sprigs hung up in a room will keep it free from flies and it is surprising that Shakespeare only referred to its extreme bitterness and did not mention this quality. In *Lucrece*, he uses the plant to signify bitterness and remorse and compares it with the sweetness of sugar: 'Thy sugar'd tongue to bitter Wormwood taste'. He frequently mentions sugar and by Elizabethan times it must have become widely used for sweetening, for its name had already become well-known in our language. Wormwood is mentioned again in *Romeo and Juliet* when the old Nurse and Lady Capulet quietly discuss Juliet's age; it was apparently the custom of the time to rub the breasts with Wormwood or aloes when weaning a child:

> Nurse: . . . I remember it well,
> 'Tis since the earthquake now eleven years;
> And she was weaned, – I never shall forget it,
> Of all the days of the year upon that day:
> For I had then laid wormwood to my dug,
> Sitting in the sun under the dove-house wall,
> My lord and you were then at Mantua: –
> Nay, I do bear a brain: – but, as I said,
> When it did taste the wormwood on the nipple
> Of my dug, and felt it bitter, pretty fool.

Again, in *Love's Labour's Lost* Shakespeare associates the word with 'bitterness' when Rosaline tells Berowne that if he wants to marry her, he will have to:

> . . . weed this wormwood from your fruitful brain
> (Without the which I am not to be won) . . .

It was one of the plants used by Girolamo Rusceli, known as *Master Alexis of Piedmont*, in his original treatise on perfumes published in 1555; to make 'a good perfume against the plague' when heated over embers 'you must take Mastich, Chypre, Incense, Mace, Wormwood, Myrrh, Aloes, Musk, Ambergris, Nutmegs, Myrtle, Bay, Rosemary, Sage, Roses, Elder, Cloves, Juniper, Rue and Pitch. All these things stamped and mixed together, you shall set upon the coals and so perfume the chamber'.

Chypre was as popular a perfume in those times as it is today and was composed of benzoin, storax, calamint, coriander seed and the dried root of *Acorus calamus*, mixed and ground to a fine powder.

SEA WORMWOOD   *A. maritima*   Pl. 23

An aromatic perennial growing 12 in. tall and resembling *A. absinthium* in that its 2-pinnatifid leaves are densely covered on both sides in silky hairs and are strongly aromatic. The flowers are minute, like those of Mugwort, and are a brownish-orange colour. There is a tendency for them to droop. In bloom July–September.

Habitat: Present in salt-marshes and around the coastline of England and Wales. Rare in Scotland and in Ireland.

SCOTTISH WORMWOOD   *A. norvegica* var. *scotica*

An almost prostrate tufted perennial growing 3–4 in. tall with deeply pinnatifid leaves covered in silky down and powerfully aromatic. The nodding flowers are rayless and are borne solitary on 3-in. stalks. They have sepal-like bracts. In bloom July–September.

Habitat: Present only on mountainous slopes in western Ross; also in Norway and Sweden.

CARDUUS   Annual or perennial herbs with leaves spirally arranged and with spiny margins. The sepal-like bracts are also spiny. Flowers hermaphrodite, unrayed, with tubular florets in which honey is secreted. Pollination by longer-tongued insects. Differs from *Cirsium* in that its pappus is unbranched. Few have scented attractions.

MUSK THISTLE   *Carduus nutans*   Pl. 24

A handsome erect biennial growing 2–3 ft tall with furrowed stems, often with spines, and lanceolate leaves deeply pinnatifid with 2–5-lobed segments, spiny at the tips. The stem-clasping leaves are covered in spiny hairs which give them a greyish appearance. The drooping flowers are borne solitary or in small corymbs and measure 1 in. across. They are royal purple in colour with a slight musky scent when inhaled near-to. The whole plant takes on a musky scent when warmed by the sun. Pollination is by bees and butterflies. In bloom June–September.

Habitat: Common on waste ground, in arable fields and in hedgerows, usually on calcareous soils throughout the British Isles, but mostly in E England and the Midlands.

History: Thistles have always been held sacred to Thor, and were believed to protect nearby buildings from the destructive effects of lightning. Thistle-down was collected by countryfolk for stuffing into cushions and pillows, usually with the dried aromatic leaves of Worm-wood and Chamomile. Thistles are prolific in their germination as one old rhyme says:

> Cut your thistles before St John,
> Or you'll have two instead of one.

Their hairy, grey appearance reminded William Blake of old age:

With my inward eye, 'tis an old may grey;
With my outward, a thistle across my way.

CIRSIUM Annual or perennial plants with spirally arranged
leaves, deeply lobed and usually spiny at the margins. Flowers borne
solitary or in clusters; florets tubular, ray-florets absent. Pappus
branched, with feathery hairs united at base.

CREEPING THISTLE *Cirsium arvense* Pl. 24
Syn: *Carduus arvensis*
A hairless perennial with a rhizomatous rootstock, growing 1–3 ft tall,
the stems without spines. The sessile leaves, hairy on the underside,
have wavy margins and are deeply lobed, ending in spines. The mauve-
pink flowers with their long tubes are borne in clusters of 3–4 and have
a sweet honey or musky scent, like *Buddleia*, which makes them
particularly attractive to butterflies. Staminate and carpellate flowers
are found on different plants which usually grow in separate groups.
In bloom June–September.
Habitat: Common on arable land, by the wayside and on wasteland
throughout the British Isles; it is a troublesome but handsome weed.

SAUSSUREA Non-spinous perennial herbs with spirally-arranged
leaves. Flowers hermaphrodite, borne solitary or in corymbs, florets
tubular, anthers with terminal tails or appendages. Pappus arranged in
2 rows, the outer as bristly hairs.

PURPLE HAWKWEED *Saussurea alpina* Pl. 24
The only British and European species, it is perennial with grooved
stems 9–12 in. tall and covered in short hairs. The lanceolate toothed
leaves are dark green above, downy on the underside, while the rayless
flowers are borne in terminal clusters with hairy sepal-like bracts. They
are purple-blue, white on the underside and have a soft almond-like
scent. In bloom August–September.
Habitat: Rare, in alpine pastures and about quarries and cliffs, usually
close to the sea and mostly confined to N England and Scotland. Rare
elsewhere, also in Ireland.

CREPIS Annual or perennial herbs with branched leafy stems
containing a milky juice. Leaves radical, spirally arranged. Flowers
borne in corymbs or panicles. Bracts many, arranged in single row.
Florets yellow, occasionally white. Fruit tapering or beaked; pappus,
many rows of simple hairs. Plants tend to be strong-smelling and are
visited by bees and hover-flies.

STINKING HAWKSBEARD *Crepis foetida*

An annual or perennial, it grows about 12 in. tall, the stems being furrowed and hairy with the leaves forming a basal rosette. They are narrow and pinnately lobed with long peduncles, and are covered in hairs. The flowers are bright yellow with sepal-like bracts, and droop in bud, opening to about 1 in. across. The whole plant has a fetid but not unpleasant smell, like that of almonds with bitter undertones. In bloom June–August.

Habitat: Found across Europe from the British Isles, through Germany and as far east as Yugoslavia, it is present in England only in Kent and Sussex and very occasionally in Cambridgeshire, growing in sandy soils by the wayside.

FLOWERING RUSH FAMILY Butomaceae

Tall aquatic rhizomatous herb with sword-like leaves. Flowers borne solitary or in umbels on cylindrical peduncle. Perianth in 2 whorls; stamens 9; carpels 6, united at base; ovules many; seeds small.

FLOWERING RUSH *Butomus umbellatus*

A rhizomatous perennial growing 3–4 ft tall with sword-like radical leaves of the same length as the cylindrical, leafless flowering stem. The flowers are rose-pink with darker veins and measure 1 in. across. They are borne on long footstalks, as many as a dozen or more forming an umbel. They emit a soft almond-like scent and are followed by purple fruits which make it one of the loveliest of all native flowers. In bloom July–September.

Habitat: In ditches and by rivers and streams, though never by stagnant water, and mostly confined to S England and Ireland.

History: It takes its botanical name from the Greek *bous*, an ox, and *temno*, to cut, for the sharp edges of the sword-like leaves would cut the mouth of cattle that grazed them. Countrymen used to inhale the slightly fragrant seeds to help them relax before going to sleep.

LILY FAMILY Liliaceae

Mostly bulbous, herbaceous plants with leaves which are never joined to the flowering stem. Flowers usually hermaphrodite with the perianth petaloid, in 2 whorls united into a tube. Stamens 6–10, in 2 whorls, opposite perianth-segments. Ovary superior, 3-chambered; style 1; stigma simple or 3-lobed. Fruit a many-seeded capsule; seeds with fleshy endosperm.

Mostly plants of the temperate regions among which several are

Plate 23
Daisy Family (i)   1 Wormwood   2 Sea Wormwood
                   3 Tansy   4 Yarrow

Plate 24
Daisy Family (ii) 1 Creeping Thistle   2 Purple Hawkweed
3 Musk Thistle

edible, e.g. Onion, Asparagus. Many bear flowers which possess the most powerful scent of all flora. This is due to the presence of indol, methyl indol being the active principle of civet, the most powerful of all animal scents. Indol is present in the early stages of putrefaction and in excreta, but in many flowers it is lightened and made more pleasant by combining with the ester, methyl anthranilate, which has the fruity scent of jasmine and of orange blossom and is present in their attars. Usually methyl anthranilate combines with benzyl acetate and indol to produce a 'heavy' scent, but in several Liliaceae, e.g. Lily of the Valley and Angular Solomon's Seal, this is tempered by pleasanter undertones.

Flowers of this family were originally greenish-white with purple anthers. Their honey was freely exposed and they either had no perfume or smelled unpleasant and were visited by flies. As the honey gradually concealed itself, the green colouring disappeared (green flowers have no scent) and white became predominant. The flowers became powerfully scented, especially at night, which was intended to draw nocturnal Lepidoptera, for the honey was now no longer accessible to the shorter-tongued insects. Several of the most beautiful and exquisitely scented of British wild flowers are members of this family.

CONVALLARIA Much-branched perennials with a creeping rhizomatous rootstock, rooting at the nodes. Leaves with long petioles, borne in pairs; scale leaves at base from which the flowers rise in a drooping raceme. Perianth globose; stamens united to middle. Berries red.

LILY OF THE VALLEY   *Convallaria majalis*   Pl. 25
A glabrous perennial with its dark green root leaves produced in pairs and with a sheathing petiole. The white bell-shaped flowers are borne 6–12 in a 6-in. tall 1–sided raceme and are deliciously scented. They are followed by globose red berries. Pollination is by bees and nocturnal Lepidoptera but in their absence self-pollination is possible, for the anthers are grouped closely around the style and discharge their pollen on to the 3-lobed stigma. In bloom May–July.

Habitat: In low-lying shady places and in deciduous woodlands where it enjoys a cool, leafy soil and the diffused rays of the sun. It is almost entirely confined to England and to those soils of a calcareous nature.

History: With its dangling white bells, tinted with green and backed by twin leaves of darkest green which display the bloom as if held in a sheath, it is one of the loveliest of all our native flowers with a spicy scent, like that of the Dame's Violet which draws the night hawkmoths from a distance. This was the flower which the poet Shelley loved so much:

> . . . the naiad-like lily of the vale,
> Whom youth makes so fair and passion so pale,
> That the light of its tremulous bells is seen
> Through their pavilions of tender green.

Its name is derived from the Latin *convallis*, a valley, for it is to be found partly hidden by the leafy trees of woodlands and vales. To the early writers it was known also as the *Wood Lily*. In the *Profitable Art of Gardening* (1568), Thomas Hyll wrote of 'the Wood Lily or Lilly of the Valley (which) is a flour marvellous sweete, flourishing in the springtime and growing properly in woods but chiefly in vallies and on the sides of hills. But now . . . is brought and planted in gardens'. Henry Lyte in his *New Herbal* (1578) calls it the *Lilly Convall* and describes the flowers as being 'as white as snow and of a pleasant strong savour. The water of the flowers comforteth the heart. The same water, as they say, doth strengthen the memorie and restoreth it again to its natural vigour, when through sickness it is diminished'. Culpeper also recommended it as a memory restorative.

In *The Flower Garden* (1726) John Lawrence described the scent of the Conval-lily as the sweetest of all, neither offensive nor over-bearing (p. 14).

Gerard tells us that it grew plentifully on Hampstead Heath (four miles from London, where it still grows in Kenwood) and at Lee, in Essex. He says that because it flowered in May (hence its name *majalis*) it was also known as the *May Lily*. In French it is known as *Muguet*.

> Sweet May Lillies richest odours shed
> Down the valley's shady bed;

wrote Sir Walter Scott. The flower was also known as *Our Lady's Tears*, as the tiny dangling bells were thought to be like large tear drops seen from a distance.

'The flowers . . . distilled with wine, . . . restoreth speech unto those that have the dumb palsie . . . they are good against the gout and comforteth the heart', wrote Gerard. In ancient times, the distilled water was held in such high esteem that it was kept only in vessels of gold. Hence Matthiolus called the water *Aqua aurea*.

The belief that the plant was able to restore the memory is substantiated by the dried flowers and roots acting like snuff when inhaled, discharging the mucous to clear the head and to bring relief to a tired mind. Ettmüller, a Leipzig physician, suggested this formula: 'of the dried flowers of Lily of the Valley and of the leaves of marjoram, a drachm each powdered. Mix well together with half a scruple of the essential oil of marjoram, and use as snuff'. The scent of the dried roots differs from that of the flowers in that it is more spicy.

The manner in which the bells are suspended from the stem gives them a lightness possessed by few other flowers. Keats said of them:

> No flower amid the garden fairer grows
> Than the sweet lily of the lowly vale,
> The Queen of Flowers.

And Bernard Barton has captured their delicate charm in his lines:

> The lily of the vale, whose virgin flower
> Trembles at every breeze beneath its leafy bower.

The plant is indigenous to almost all parts of Europe from Britain to the Caucasus and grows freely in the woods of France and Germany where it is known as *Meyen Blumen*. In the garden, it requires the same conditions as in its native woodlands, either a situation facing north or beneath trees or shrubs to provide partial shade. But it must have dampness about its roots and this is encouraged by working humus into the soil in the form of peat, leaf-mould or used hops, which will retain much of the winter's moisture during spring and summer when it is in bloom.

Specially retarded crowns (roots) are available from nurserymen in Germany for forcing in gentle heat in pots, to be taken indoors for winter flowering when their fragrance is always appreciated.

There is a double form, *flore plena*, and another bearing pink flowers, *rosea*, but neither is as lovely as the single white form with its highly scented flowers. A small shaded bed, with the fragrant pink Polyanthus Enchantress planted about it, presents a delightful picture of pink and white in May and June. The two plants are very well suited to accompany each other in small vases in the home. Hartley Coleridge, contemporary of Shelley, considered the Lily of the Valley to be the loveliest of all scented flowers. His lines dedicated to the flower are among the most beautiful in our language:

> Some flowers there are that rear their heads on high,
> The gorgeous products of a burning sky,
> That rush upon the eye with garish bloom,
> And make the senses drunk with high perfume.
> Not such art thou, sweet Lily of the Vale!
> So lovely, small, and delicately pale,
> We might believe, if such fond faith were ours,
> As sees humanity in trees and flowers,
> That thou wert once a maiden, meek and good,
> That pined away beneath her native wood
> For every fear of her own loveliness,
> And died of love she never would confess.

POLYGONATUM Herbaceous perennials of the northern hemisphere with a creeping rhizomatous rootstock from which flowering stems rise 2–3 ft tall and leave 'seals' or marks on the roots. Leaves cauline; flowers borne solitary or in terminal or axillary racemes. Perianth 6-cleft, tubular, united; stamens 6, inserted in tube; anthers 2-lobed; style with 3-lobed stigma. Fruit, a berry. Pollination mostly by bees, while self-pollination is possible as in *Convallaria*.

ANGULAR OR SCENTED SOLOMON'S SEAL
Polygonatum odoratum   Pl. 26
Syn: *P. officinale*
A perennial with a graceful arching stem 12–15 in. long, differing from other species in its angular stem and scented flowers. The leaves are thick, dark green, ovate and sessile, and are borne at regular intervals along the stem. The greenish-white flowers are tubular and are borne 1 or 2 on a common peduncle. They have a powerful sweet scent and are followed by black globose fruits. In bloom June–July.
Habitat: In woodlands of N England and west of the Pennines; also in Wales, usually in a calcareous soil. Very rare in Scotland; absent from Ireland.
History: Because of its linear verticillate leaves growing in the form of a ladder, countrymen have named the Common Solomon's Seal, *Ladder-to-Heaven*. There is a general belief that the name *Seal* was given to the plant because the markings of the fleshy stems on the roots leave the impression of Solomon's seal; but more credible is the theory of Dioscorides, who said that the roots of the plant, when dried and pounded and laid on wounds, healed (or sealed) them in the quickest possible time. A medical author of the reign of Elizabeth I wrote that 'the roots of Solomon's Seal, stamped and applied whilst fresh and green taketh away in one night any bruise, black or blew spots gotten by falls or woman's wilfulness in stumbling upon their hasty husband's fists'.

The ancient name *Solomon's Heal* became with the passing of time *Solomon's Seal*.

The three British species are delightful plants of the woodlands, requiring similar conditions to the Lily of the Valley. They grow taller, however, varying in height from 1–3 ft, and tend to flower later. Miss Sinclair Rohde has described the fragrance of the flowers as 'a curious rich "thick" smell, quite unlike that of any other English flower', and it is indeed like the scent of the Tuberose – heavy but agreeable. She did, however, fail to say that it is only the Angular Solomon's Seal that has scented flowers, perfume being completely absent from the common form.

The bells, like those of the Lily of the Valley, were in great demand during Elizabethan times for distilling; the water was used to remove

Plate 25
Lily Family (i)   1 Scented Solomon's Seal 2 Lily of the Valley

Plate 26
Lily Family (ii)   1 Wild Tulip   2 Nodding Star of Bethlehem
                       3 Bluebell   4 Spring Squill

freckles 'leaving the face fresh, fair and lovely'. Parkinson tells us that it was a favourite toilet 'wash' among Italian women.

LILIUM   Perennials with scaly bulbs and erect leafy stems. Leaves cauline, alternate or whorled. Flowers borne solitary or in loose racemes, drooping, horizontal or erect. Perianth funnel- or bell-shaped with 6 segments. Stamens 6, with long slender filaments; anthers brown, red or orange. Fruit a 3-celled and 3-valved capsule.

TURKSCAP LILY   *Lilium martagon*

A perennial growing 2–3 ft tall from bright yellow bulbs and bearing whorls of glossy dark green lanceolate leaves, 6–9 in each whorl. The drooping flowers are borne in erect chandelier-like racemes of 20 or more, are dull purple heavily spotted with carmine and have recurved perianth segments. They have a heavy perfume and are pollinated by butterflies and by nocturnal Lepidoptera. In bloom July–August.

Habitat: A garden plant since Tudor times, it is to be found in several woodland areas in the West Country, also in Gloucestershire and Monmouthshire.

History: Bulbs may first have reached England towards the end of the sixteenth century, for this is possibly the plant referred to by Gerard when he said 'this plant groweth in the fields and mountains, many days journey beyond Constantinople, whither it is brought by the poor peasants to be sold for the decking up of gardens. From whence it was sent with many bulbs of rare and dainty flowers by Master Harbran, our ambassador there, unto my honourable good lord and master the Lord Treasurer of England, who bestowed them upon me for my garden'. The plant may, however, be a rare native, for Dr. Turner, who was Dean of Wells and dedicated his *Herbal* to Queen Elizabeth I, mentioned a Lily of reddish-purple colouring. As it is found in several woods in Somerset, as well as in Devon and Dorset, he may have seen it growing wild, though its presence there was unknown to Gerard. Parkinson grew several varieties, including the lovely white *album* and tells us that he counted 'three score blooms growing thick together on one plant'. Bulbs may have reached England with returning Crusaders, for during early times it was known as the *Lily of Nazareth*.

FRITILLARIA   Perennials with scaly bulbs covered in a white tunic and with erect leafy stems. Flowers borne solitary or in pairs. Perianth bell-shaped, drooping; segments not recurving as in *Lilium*, with nectary at base of each segment. Style with 3 stigmas. Pollination by Hymenoptera but self-pollination possible. Crown Imperial, *F. imperialis*, has a foxy or honey-like smell; *F. meleagris* has a soft mossy fragrance which is more pronounced in the white forms.

SNAKE'S-HEAD FRITILLARY, GINNY-HEN FLOWER *Fritillaria meleagris*

It has small scaly bulbs from which 9-in. stems rise with grey-green, grass-like leaves. Solitary, nodding, bell-shaped flowers are borne at the end of the stems, and dangle in the early summer breezes. The flowers are usually dull purple, occasionally white with darker markings. In bloom April–June.

Habitat: In moist meadows on either side of the Thames where it grows in large colonies; especially conspicuous in Berkshire and Oxfordshire.

History: It takes its name *meleagris* from the Latin for guinea-fowl, because of the dull purple-grey colour of its flowers and its unusual markings which, as Gerard said, resemble those of the guinea-fowl or *Ginny-hen*. He also named it the *Checkered Daffodil* and tells us that it was introduced during the reign of Elizabeth I. Gerard himself received plants sent by one Jean Robin of Paris. In France the plants are found in abundance and 'were greatly esteemed for the beautifying of gardens and the bosoms of the beautiful'. Parkinson also knew it as the *Checkered Daffodil* and says that it was named *Narcissus caparonius* in honour of Noel Capron, an apothecary of Orleans who was murdered in the massacre of St Bartholomew's Day.

From several authorities we learn that the bulbs were brought from France, together with the Auricula, by French refugees fleeing from the massacres, who set up the silk trade first in Canterbury, later at Spitalfields. But whereas the Auricula has remained a garden plant, the Fritillary has for long been naturalised along the banks of the Thames, from which it would appear to be a native flower. Those known to Gerard and Parkinson may simply have been different varieties. In France, the Black Fritillary is common and Henry Phillips mentions that a pure white form is also to be found in the fields around Poitou. The checkered White Fritillary is found in Berkshire meadows by the Thames, but not the pure white form. The name *Fritillary* signifies a dice-box and is derived from the shape of the fully opened blooms which are almost square, like lanterns, with buds greatly resembling the head of a snake; hence its country name.

Matthew Arnold knew where it could be found, as he tells us in *Thyrsis*:

> I know what white, what purple fritillaries
> The grassy harvest of the river fields
> Above by Eynsham, down by Sandford, yields
> And what sedged brooks are Thames's tributaries.

The first to find it growing wild in England – a Mr Blackstone – said he had seen it in Maude Fields, near Ruislip. Shortly afterwards it was found first at Mortlake and at Kew, and then near Oxford.

TULIPA   Perennials with bulbs covered in a brown tunic and radical, cauline leaves. Flowers borne solitary or in twos with a bell-shaped perianth of 6 segments; nectaries (as in Fritillary) absent. Style absent, stigma 3-lobed. Seed-capsule flat, erect. Many garden varieties are scented and are mostly derived from *T. gesneriana* which bears fragrant scarlet flowers and is native to C Asia and Asia Minor. It is one of the few scarlet flowers with perfume. *T. suaveolens*, which is native to the same regions and is the ancestor of the dwarf Duc van Thol Tulips used for Christmas forcing, is also scented, as is our native *T. sylvestris*. Pollination is by almost all forms of insect life.

WILD TULIP   *Tulipa sylvestris*   Pl. 26
A glabrous perennial with a brown tunicated bulb producing a round stem about 12 in. tall. It has glaucous lanceolate leaves near the base which fold over at the tips. The flowers, which droop in the bud, open wide and star-like with long pointed petals and filaments which are hairy at the base. They are pale yellow, shaded green on the outside and have a soft, sweet scent. In bloom April–June.
Habitat: In chalk-pits and meadows of calcareous soils; also in woodlands and orchards, mostly in Somerset and Gloucestershire; SE England; SE Scotland.

ORNITHOGALUM   Glabrous   perennials   with   linear   glaucous leaves and flowers borne in a bracteate or corymbose raceme. Perianth of 6 free spreading segments, each with a nectar-gland at the base. Filaments lanceolate; anthers versatile; flowers white. Pollination by small insects or self-pollinated, *O. umbellatum* being slightly perfumed and *O. arabaicum*, a native of the Middle East, having a more pronounced scent.

NODDING STAR OF BETHLEHEM   *Ornithogalum umbellatum*   Pl. 26
A glabrous perennial growing 6–8 in. tall with dark green linear leaves which have a central stripe of white. The flowers are borne in a corymbose raceme of 6–10 and are chalky white with a central stripe of pale green down each petal. They open star-like, resembling *Tulipa sylvestris*, and have a soft cucumber-like perfume. In bloom May–June.
Habitat: On grassy banks and in meadows, usually growing in a sandy soil and mostly confined to England. Rare elsewhere.
History: To the countryman the flower is known as *Eleven-o'clock Lady* for whatever the weather the blooms do not open until one hour before noon and close again at three o'clock. For this reason the flowers remain fresher than any others, not only in the wild but also when they are cut and taken indoors. On the plant they will retain their full beauty for 3–4 weeks. Henry Phillips said that 'the flowers have a pleasant

odour' and tells us that countrymen used to roast and eat the nourishing bulbs. The plant is found right across Europe, and Phillips said 'we can have no hesitation in pronouncing it indigenous to our soil' for it had been found growing wild in Surrey, Norfolk and Cambridgeshire, in Christ Church meadows at Oxford, and near Knaresborough and Leeds in Yorkshire.

SCILLA Glabrous perennials, with bulbs covered in a brown tunic. Leaves linear, usually appearing before the flowers which are borne in a scape or raceme. Perianth-segments purple, blue or white, free, with mid-rib. Filaments inserted at base of perianth. Fruit a 3-chambered, multi-seeded capsule. Pollination by Diptera, attracted to the nectar exposed in a shallow cup as the flowers open, by the purple anthers and slight indol smell.

SPRING SQUILL *Scilla verna* Pl. 26
It grows 4–6 in. tall with grass-like root leaves and has leafless stalks bearing at the end corymbs of deep blue scented flowers which open star-like and have extended lanceolate bracts and violet anthers. In bloom April–June.
Habitat: On grassy banks and cliffs, usually close to the sea and mostly in Cornwall; also in NW England and Scotland. Present on the E coast of Ireland, especially around Wexford.

HYACINTHOIDES Glabrous perennials differing mainly from *Scilla* in that they renew their bulb annually. Leaves linear, produced before the bracteate flower-scape. Perianth campanulate, segments united at base. Stamens inserted at middle of perianth.

BLUEBELL, WILD HYACINTH *Hyacinthoides non-scripta*
Pl. 26 Syn: *Endymion non-scriptus, Scilla non-scripta, S. pestalis, S. nutans, Hyacinthus non-scriptus*
A glabrous plant with linear leaves 12–15 in. long and bearing a round leafless stem of similar height. At the end is borne a bracteate raceme of 6–12 drooping bell-shaped flowers of purple-blue which have the balsamic perfume of the Oriental Hyacinth. The perianth segments turn back attractively at the tips and expose the white anthers. Pollination is mostly by Hymenoptera. In bloom May–July.
Habitat: Common in woodlands and in fields and hedgerows throughout the British Isles, preferring a soil of slight acidity and partial shade.
History: It has been given more botanical names than any other plant and has from earliest times been confused with the Harebell of the moorlands – the Bluebell of Scotland. Robert Burns wrote of 'hyacinth for constancy with its unchanging blue'. Gerard knew it as the *English*

Jacinth or *Blue Harebell* and describes the flowers as having 'a strong sweet smell, somewhat stuffing the head'. The scent is spicy, like balsam or cinnamon, for cinnamic alcohol is present in the flowers.

From the pear-shaped bulbs, about 1 in. in diameter, the Elizabethans obtained a starch to stiffen their ruffs, and glued together the pages of their books with the sticky substance secreted by the flower-stalks. Gerard said that it was also used for sticking feathers on arrows. The poet Shelley sings its praises in his lines to 'The Sensitive Plant':

> And the hyacinth purple, white and blue
> Which flung from its bells a sweet peel anew
> Of music so delicate, soft and intense,
> It was felt like an odour within the sense.

ALLIUM Unbranched perennials, bulbous or rhizomatous. Leaves radical with sheathing stem at their base. Flowers borne in an umbel with a 1–2-leaved spathe and perianth of 6 free segments. Filaments dimorphic, the inner one bearing the anther. Fruit a 3-chambered capsule. Almost all species have a strong onion-like smell due to the presence of sulphur compounds in varying degrees of strength. Pollination usually by Diptera of all types, though self-pollination is usually possible in their absence.

CHIVES *Allium schoenoprasum*

Syn: *A. sibiricum*

A tuft-forming perennial without bulbils and growing 6–8 in. tall. The grey-green leaves are round and hollow and release a soft onion-like smell when broken. The flowers are purple-pink and are borne in small round heads, like Thrift, each floret having a paper-like bract and undivided stamens. In bloom June–August.

Habitat: A rare plant of rocky limestone pastures, mostly of SW England; Northumberland and NE Yorkshire; S Wales and the West coast of Ireland.

History: The plant was mentioned in an early fifteenth-century manuscript of cooking under its old name of *seives* as well as in *Britannia's Pastorals*, in the account of King Oberon's feast:

> Straightways follow'd in
> A case of small musicians, with a din
> of little Hautbois, whereon each one strives
> to show his skill; they all were made of seives
> excepting one . . .

The chopped leaves are used in omelettes and salads to impart their mild piquant flavour.

RAMSONS *A. ursinum*

A broad-leaved species, the bright green leaves resembling Lily of the Valley but releasing an obnoxious fetid smell of stale garlic. This is released with the slightest movement, for example a gentle breeze, so that quite a small colony may be detected from a distance. The flowers are white and are produced in a flat inflorescence on a 3-angled stem 12–14 in. tall. In bloom April–June.

Habitat: Usually in damp woodlands and hedgerows, especially in SW England where colonies are particularly prominent around the Roman city of Bath.

History: To countrymen it is known as the *Broad-leaf Garlic*, though its botanical name is taken from the similarity of its leaves to a bear's ear. The whole plant (including the root) gives off a fetid sulphurous smell. The leaves are used in the West Country to flavour pilchards and winter stews.

HERB PARIS FAMILY Trilliaceae

Rhizomatous perennial herbs with a simple erect stem. Leaves net-veined, opposite or borne 4 or more in a whorl. Flowers borne solitary or in a terminal umbel; usually green. Segments unequal. Stamen equal in number to, and opposite, perianth segments. Ovary superior; single-celled. Fruit, a berry or fleshy capsule.

PARIS   Erect perennial plants with a creeping rhizomatous root-stock. Leaves reticulate-veined, borne in whorls of 4 or more. Flowers borne solitary with narrow outer petaloids. Filaments short; anthers elongated. Fruit, a black fleshy capsule.

HERB PARIS *Paris quadrifolia*

A glabrous perennial growing 12–14 in. tall, the stem leafless except for a whorl of 4 netted leaves at the top. At the centre appears the green flower which has 4 narrow sepals and 4 narrow petals, all in fours, hence its name. With its obnoxious smell and purple stigmas which glisten as if covered with moisture, it attracts carrion-feeding Diptera which alight on the stigma and climb up the anthers, covering their feet with pollen before flying off to other flowers. In bloom May–July.

Habitat: In dark deciduous woodlands throughout the British Isles but not in Ireland, usually growing in colonies and on calcareous soils.

DAFFODIL FAMILY Amaryllidaceae

Bulbous herbs with sword-like radical leaves. Flowers solitary or cymose, distinguished from those of the Liliaceae by their inferior ovary. Flowers, before opening, enclosed in spathe. Perianth in 2 whorls, often with corona. Stamens 6, borne in 2 whorls, opposite perianth segments, with versatile anthers. Ovary inferior, 3-chambered. Style 1; stigma 3-lobed. Fruit a many-seeded capsule; seeds with endosperm.

LEUCOJUM Bulbous perennials, differing from *Galanthus* in having more than 2 leaves and perianth segments alike. Leaves linear, blunt, glaucous or brilliant green. Flowers nodding, borne solitary or 2–6. Perianth segments white, tipped green, thickening at tips. Stamens 6, equal. Corona absent. Fruit, a capsule; seeds black.

Of the two British species the Spring Snowflake, *L. vernum*, needs its scent to attract the few insects about when it is flowering and is pollinated by Lepidoptera. The Summer Snowflake *L. aestivum*, is pollinated by Hymenoptera, and therefore does not need to be scented for its reproduction.

SPRING SNOWFLAKE *Leucojum vernum*

It forms a large green globose bulb and grows 6–8 in. tall with 3–4 broad linear bright green leaves and bears its nodding flowers 1 or 2 together. The pointed petals of ivory-white are tipped with green, while the orange anthers are conspicuous. Fragrant at night, it is visited both by day- and night-flying Lepidoptera. In bloom March–May.

Habitat: Very rare, though present on certain hilly pastures in Somerset, Dorset and Wiltshire, usually where there is some shade.

History: The scent of the flowers has been compared to that of the Hawthorn's, but does not have the same fishy undertones. The plant takes its name from the Greek *leukos*, white, and *ion*, violet, for its scent resembles that of the Violet. So powerful is its perfume that Gerard classed it with the Gillyflowers, naming it the *Bulb Stock Gillyflower*. Its scent was thought to be similar to that of the Stock and Wallflower and is very like the Mignonette's; these are all flowers of the 'Violet group' and Gerard was accurate in placing them together.

It is a delightful but rarely seen bulb for the alpine garden, and for cool greenhouses or frames, and requires a well drained gritty compost.

GALANTHUS Bulbous hairless perennials with 2 radical leaves and a flattened peduncle. Flowers solitary, drooping. Perianth of 6 free segments, the outer segments larger and spreading. Corona absent; tube short. Capsule ovoid.

SNOWDROP, CANDLEMASS BELLS, FEBRUARY FAIR-MAIDS *Galanthus nivalis*

From the small ovoid bulbs arise a pair of grooved grey-green leaves. The spathe is colourless with 2 green veins. The solitary drooping flowers are borne on 6-in. stems and have 3 white spreading outer petals and 3 bell-shaped inner petals, tipped with green, also shaded green on the outside. In bloom January–March.

Habitat: The plant inhabits the coolest regions, and is naturalised in Moray and Fife, in Northumberland and N Yorkshire, usually growing in damp deciduous woodlands. The plant is also found in the Pyrenees, the alpine regions of Italy; in Greece and Bulgaria; the Crimea and Caucasus. Sir Reginald Dorman-Smith of Tresco has said that *G. nivalis* will not grow on the Isles of Scilly as the climate is too dry.

History: To distinguish it from the Snowflake with which it was always associated and confused, Linnaeus gave the Snowdrop its botanical name *Galanthus*, or *milk-flower*. It is one of the best loved of all wild flowers as it is the first to herald the awakening of the countryside after the dull days of early winter. As the poet Thomson said:

> Fair-handed spring unbosoms every grace,
> Throws out the Snowdrop and the Crocus first.

William Wordsworth in his lines 'To a Snowdrop' wrote:

> Chaste Snowdrop, venturous harbinger of spring
> And pensive monitor of fleeting years!

G. nivalis is native of Italy and was known to St Francis. It may be a native flower or it may have been brought to Britain by monks during the fifteenth century, for it can be found on the sites of old monastery gardens everywhere. It was planted to provide blooms for Candlemass Day, the Feast of the Purification of the Virgin on the first day of February, hence its delightful country names of *Candlemass Bells* and *February Fair-maids*.

The early poets do not mention it and Gerard knew it only by the name of *Early-flowering Bulbous Violet* and classified it with the Snowflakes. Shakespeare makes no mention of it, from which we may assume that it is not a native plant. We first hear of its name, the *Snowdrop*, from Johnson who revised and published Gerard's work as a new edition in 1633. The name seems to have been preferred to all others. In Germany it is given the equally charming name of *Snowbell*.

The flower has been specially equipped by nature to withstand cold winds and nights of severe frost. The manner in which the corolla is attached to the flower stalk enables it to be blown about by strong winds without fear of its breaking away, while its drooping position enables it to throw off moisture, thus protecting the pollen. Also, the whiteness of the petals acts as a reflector which attracts the warmth of the daytime sunshine, however little this may be, to the anthers. This helps to perfect the pollen. The flowers also tend to close up at night, and thus retain the maximum warmth of the sunlight during the hours of darkness.

Like all the Snowdrops, *G. nivalis* grows in Greece and on her islands, and Theophrastus described it as being visited by bees on Mount Hymettus. When it visits the flowers in sunny weather, the hive-bee alights on the outer petals. It then pushes its head and forelegs into the flower, clinging on to an inner perianth segment with its hindlegs. Then with the tarsal brushes of its forelegs it sweeps pollen from the anthers into the baskets on its hindlegs. The style projects beyond the anthers so that the bee touches the stigma when visiting another flower into which it thrusts its pollen-covered head. In absence of bees, self-pollination is made possible as the pollen falls on to the stigma. The newly opened flowers have a soft honey-like scent which is most pronounced where they grow in large numbers or indoors in a warm room. The scent is mentioned by Clusius and is more noticeable in several hybrid forms such as Arnott's Seedling.

NARCISSUS Bulbous perennials with linear glaucous leaves and a membranous spathe. The white or yellow flowers are borne solitary or in umbels, drooping and horizontal. The perianth has 6 equal petaloid segments in 2 whorls, united into a tube below, the mouth of which terminates into a bell-shaped corona. Stamens 6, with versatile anthers. Pollination of the yellow early-flowering *N. pseudonarcissus* is usually by bees; the introduced *N. poeticus* with its white flowers is visited by nocturnal Lepidoptera.

WILD DAFFODIL, LENT LILY *Narcissus pseudonarcissus*
It has blunt glaucous strap-like leaves 9 in. long and a hollow stem 9–10 in. tall, at the end of which appears a solitary yellow flower, with a long trumpet notched at the margin and protruding beyond the paler perianth petals. In bloom March–May.
Habitat: In deciduous woodlands and damp meadows, usually at the edges of woods, throughout the British Isles but especially in the western half of England and in Wales.
History: It may have been introduced by the Romans but was so well established early in our history that it may be a native plant. It is one of the first spring flowers to bloom and was therefore a favourite of the poets. Shakespeare, in *A Winter's Tale*, wrote these well-known lines:

> Daffodils,
> That come before the swallow dares, and take
> The winds of March with beauty.

Sir Aubrey de Vere in his 'Ode to a Daffodil' expressed a similar feeling when he wrote, 'O love-star of the unbeloved March . . .'. William

Wordsworth from his cottage home in the Lake District described his joy at finding the first Daffodils of springtime:

> I wander'd lonely as a cloud
> That floats on high o'er vales and hills,
> When all at once I saw a crowd,
> A host, of golden daffodils . . .

During Elizabethan times, the Lent Lily, so called because it was usually in bloom during Lent, grew in such abundance in the fields near the City of London that the women of Cheapside used to carry the blooms into the city every day in large circular baskets on their heads, to be sold for a few pence a bunch. They must have been familiar to Shakespeare during his stay in the Silver Street lodging-house, and would be a happy reminder of the Warwickshire countryside.

Lyte in his *New Herbal* called it the *Yellow Crowbell* or *Bastard Narcissus* for, unlike today, it was only the short-cupped species that were then called Daffodils. To Gerard and to the poet Spenser, the native form was known as *Daffodowndillie*:

> Strew me the ground with daffodowndillies,
> With cowslips and kingcups, and lovèd lilies.

Spenser's 'lovèd lilies' were Lilies of the Valley, usually in bloom at the same time.

Seeming to emit little scent when growing in the open, a bunch of Lent Lilies emit a soft, sweet, mossy perfume when indoors, rather like wild Primroses, and the larger their number the more pronounced is their fragrance.

PHEASANT'S EYE, POET'S NARCISSUS *N. poeticus*

Syn: *N. majalis*

It has broad erect leaves and bears on a 12-in. stem a solitary flower with a flat white perianth over 1-in. wide, slightly reflexed at the margins in *recurvus* and with a small yellow cup, margined with red to resemble the eye of a pheasant. In bloom April–May.

Habitat: A rare garden escape, naturalised in parts of Norfolk and SE England.

History: Native of Italy and S France, it may have been introduced by the Romans for it was naturalised at an early date and grew in most gardens of recorded history. The flowers contain indol and their extremely heavy scent prompted Burbidge to write that in a closed room it is 'extremely disagreeable, if not injurious to delicate persons'. Pliny said that the plant was named *Narcissus* from the Greek *narke*, a word denoting dullness of the senses, from which the word *narcotic* is derived. Sophocles said that the departed 'dulled with death, should be crowned with dulling flowers'. For this reason, the early Egyptians used the flowers in funeral wreaths, for *N. poeticus* bloomed in abundance in their country.

Lyte said, 'these pleasant flowers are called in English, White Daffodils and Primrose Peerless' on account of their glistening whiteness. They were also known as *Sweet Nancies*. Henry Phillips said that, 'this narcissus seldom produces seed in England even by the assistance of cultivation, and we are therefore of the opinion that plants found at Shorne, between Gravesend and Rochester, as well as those discovered in Norfolk are the offsets of imported plants, probably as of early a date as the Romans, who would not fail to plant the flower of their favourite poet'.

John Addington Symonds in his lines 'To A Narcissus' wrote:

> O blooming white narcissus-bud that lendest
> New beauty to the meadow where thou bendest!
> The spring without thy scent were nought,
> Scarce worth one thought,
> 'Mid all its paradise of buds . . .

IRIS FAMILY Iridaceae

Herbaceous plants with rhizomatous roots and sword-like leaves, usually arranged in 2 ranks. Flowers borne in a terminal cyme. Perianth superior, petaloid, of 6 leaves in alternating whorls, united below into a long or short tube. Stamens 3 with extrorse anthers; ovary inferior; 3-chambered. Style 1, stigmas 3. Fruit a 3-sided, many-seeded capsule.

Mainly plants of tropical America and S Africa, pollinated by humming-birds. Very few have scented attractions though *I. florentina* provides the violet-smelling orris root of commerce, the scent being released when the roots become dry. Several species of *Crocus* bear scented flowers including *C. vernus*, long naturalised in the British Isles. The dried scented stigma of *C. sativus* (Saffron) has been grown and used in England since medieval times. The genus takes its name from the Greek for Saffron.

IRIS Herbaceous plants with rhizomatous roots and radical sword-like leaves. Flowers with membranous spathe; perianth-tube short, sepals 3, petaloid, reflexed. Petals 3, erect; style dividing into 3 petal-like lobes covering the stamens which are inserted at base of outer perianth-segments. European species pollinated by long-tongued hover-flies, *Rhingia rostrata*, and by bees with a 7–8 mm proboscis which enter the flower in search of honey secreted at the base of the perianth.

ROAST-BEEF PLANT, GLADDON, STINKING IRIS *Iris foetidissima*

An evergreen glabrous perennial of tufted habit, growing 2 ft tall, the stem having a sharp ridge. The leaves are equal in length to the scape and when bruised release a strange though not unpleasant smell, sweetly acrid, and like that of beef roasting on a spit. The small beardless flowers are an unusual shade of yellowish purple-brown with darker veins, like well cooked meat in appearance as well as smell. They are followed by brilliant green fruits 2 in. long and 3-sided, which open when ripe to reveal bright red seeds. They may be used for winter decoration indoors though will persist on the plant throughout the winter. In bloom June–August.

Habitat: In deciduous woodlands and scrubland, usually over limestone formations and mostly confined to S England. A form bearing flowers of purer yellow is present in parts of Dorset and Hampshire and on the Isle of Wight.

YELLOW FLAG, WATER IRIS *I. pseudacorus* Pl. 27

A rhizomatous aquatic perennial with broad sword-like leaves which sends up golden-yellow flowers on a much-branched stem 3–5 ft tall. Professor Hermann Müller found that there are two forms of the flower, one in which the petaloid-styles stand close to the perianth segments and which are visited by hover-flies, and another in which the style lies some distance away from the perianth-segments. Here, the larger Hymenoptera creep under the style to reach the honey secretions and on their way touch both stigma and anther with their backs, which is not possible with smaller bees. The flowers, which measure about 3 in. across are not bearded but have the same purple veins as *I. foetidissima* and an orange spot near the base of each petal. Like most honey-secreting flowers visited by bees, they have a soft sweet honey-like scent. In bloom May–August.

Habitat: In marshland and low-lying ground and on the banks of rivers and ponds throughout the British Isles.

History: Its flower is the Fleur-de-lis of France, the 'flower of Louis', for it was taken by Louis VII as his symbol during the Crusades. It is the 'vagabond flag' of Shakespeare's *Antony and Cleopatra* and became the emblem of medieval France, in heraldic language, 'Azure powdered with fleur-de-lis or'.

A delicately scented oil was at one time extracted from the dried rhizomes, which was used as a substitute for, or to adulterate, the essential oil of *Acorus calamus*, from which it derives its botanical name.

In 1339, when Edward III made claim to the throne of France and began hostilities against Philip VI (first of the Valois kings) he took for his arms the three Plantagenet lions for two of the quarters and the Fleur-de-lis of France for the others. The great victories of the Black Prince at Crecy and Poitiers followed and were consolidated by the victory of Henry V at Agincourt. Yet within the lifetime of Henry's son, Henry VI, who had married Margaret of Anjou in 1445, England lost every French possession with the exception of Calais. In the opening scene of *Henry VI, Part 1*, the body of Henry V lies in state in Westminster Abbey, attended by the Dukes of Bedford, Gloucester, Exeter and the Earl of Warwick. A messenger arrives with the news of the losses in France:

> Sad tidings bring I to you out of France,
> Of loss, of slaughter, and discomforture:
> Guienne, Champagne, Rheims, Orleans,
> Paris, Guysors, Poitiers, are all quite lost. . . .
> Awake, awake, English nobility!
> Let not sloth dim your honours, new-begot;
> Cropp'd are the flower-de-luces in your arms;
> Of England's coat, one half is cut away.

These losses were due almost entirely to the exploits of Joan of Arc, and just before her martyrdom in 1431 Charles VII of France granted letters of nobility and the name *Du Lis* to Joan's brothers and the arms, 'Azure, a sword in pale proper, hilted and supporting on its point an open crown or, between two fleurs-de-lis'.

During Shakespeare's time the Water Iris was classed as a Lily; for example, in *A Winter's Tale* Perdita speaks of 'lilies of all kinds, the flower-de-luce being one'. The others known to Elizabethan gardeners were the Crown Imperial, the Madonna Lily and the Scarlet Turkscap, all plants of the near East introduced into England at the time of the Crusades.

CROCUS Plants with fleshy corms covered with scales; buds are formed from the axils giving rise to new corms. The corms are covered in a fibrous tunic. The linear grass-like leaves have a white midrib. Flowers usually solitary or borne in small cymes; funnel-shaped with long perianth-tube. Style slender; stigmatic lobes 3. Honey is secreted by the ovary and in several species is available only to the long-tongued Lepidoptera attracted by the sweet scent and by the white or mauve-pink colour of the flower.

SPRING CROCUS *Crocus vernus*
Syn: *C. purpureus*
Its corm is globose with a reticulated tunic and its long dark green leaves appear with the flowers. The flowers are white or mauve-pink with a long narrow tube which is almost filled by the style. The honey

secreted by the ovary rises in the tube but is available only to the longest-tongued insects. Those with white flowers release a sweet scent at night and are visited by hawkmoths, while those bearing purple-pink flowers, scented by day, attract butterflies and occasionally bees.

Habitat: Rare in meadows on high ground, in England, Wales and S Scotland.

ORCHID FAMILY Orchidaceae

Cosmopolitan perennial herbs differing widely in habit, depending on the conditions in which they grow. Some are terrestrial, others epiphytic with aerial roots, and while some inhabit the temperate regions, others are native to the tropical forests. The highly specialised flowers ensure their pollination by all types of insects – by night-flying Lepidoptera, Hymenoptera and Diptera – and though some are fertilised by their own pollen, others are sterile to their own but may be fertile to that of other species of the same genus. Among the most heavily scented of all flowers, Orchids have a highly developed scent which often varies by day and night, while the flower may take on a completely different smell after fertilisation. For example, the delicious vanilla-like scent of *Orchis mascula* before fertilisation becomes an unpleasant cat-like smell when pollination has taken place. The Pyramid Orchid, *Anacamptis pyramidalis*, which attracts day-time butterflies with its sweet perfume, at night emits an unpleasant fox-like smell to repel night-flying Lepidoptera, but becomes sweet again at daybreak. Many tropical Orchids are scentless and are visited by humming-birds which are alone able to reach the honey with their long beaks and are able to hover above the flowers for long periods as if motionless. Others are unpleasantly scented and attract hover-flies which suck out the nectar in a similar way.

Leaves entire, usually spirally arranged, often with sheathing base, occasionally spotted; in saprophytic forms, reduced to scales, without chlorophyll. Inflorescence, a spike or raceme. Flowers mostly zygomorphic; perianth of 6 segments in 2 whorls, usually petaloid or with outer whorl sepaloid, inner petaloid. British species usually provided with two rounded or palmate root tubers and with two or more basal sheathing leaves. Flowers borne in simple raceme with 3 sepals and 3 petals, the lower often spurred. The lower lip or labellum sometimes resembles a bee or other insect, occasionally a reptile. Stamens united to style to form a central column, with usually only one producing pollen. Ovary inferior, often twisted; single-chambered. Stigma hollow. Fruit a 3-valved, many-seeded capsule. Seed is smaller than that of any other plant; a single capsule of certain species is estimated to contain nearly 2 million dust-like seeds which are distributed by wind. Endosperm absent.

Of the scented Orchids, only *Vanilla planifolia* has commercial value, the dried pods yielding vanillin, which has a sweet scent and is used in the flavouring of foods and drinks.

CYPRIPEDIUM Leafy terrestrial plants with creeping rhizomatous roots. Perianth segments spreading; labellum shoe-shaped; spur absent. Column partly enclosed in labellum. Under the staminode are two anthers, beneath which is the stigma. Pollination is by small bees of species *Andrena* and by beetles (*Meligethes*). Attracted by the rose-like scent, these enter the slipper-shaped lip, the floor of which is covered with drops of honey. Then, creeping beneath the stigma, they escape through two tiny holes in the lip, covering themselves with sticky pollen grains (which are separate and not joined together as in other species) from the anthers above. The same beetles which pollinate *Cypripedium* also visit rose blossom and both plants seem to have a similar fruity scent, like that of other flowers pollinated by beetles.

LADY'S SLIPPER
Cypripedium calceolus

One of the rarest and most beautiful of Britain's wild flora. It has a creeping rhizomatous rootstock and a downy stem about 12 in. tall. The broad pale green leaves are heavily ribbed and are borne 3 or 4 to a stem. The flowers have reddish sepals and a large inflated pale yellow lip which, in addition to the scent, may be an attraction to bees and beetles. The lip is shaped like a slipper and marked with crimson on the inside. When newly open, it emits a soft sweet rose-like scent. In bloom May–June.

Habitat: Now almost extinct except in one or two deciduous woodlands in N Yorkshire and Durham, growing in a limestone soil.

Lady's Slipper

EPIPACTIS Perennial herbs with short horizontal rhizomatous roots and leaves arranged spirally or in 2 ranks. Flowers almost bell-shaped and drooping, from a distance like Lily of the Valley, with twisted pedicels. Flowers vary in number from 10–50 to a spike and in colour from greenish-white to purple-red. Pollination is usually by wasps and small bees. Only one species has scented flowers; several are self-pollinating.

RED HELLEBORINE   *Epipactis atrorubens*   Pl. 30

It has thin spreading roots, like Mint, from which several flowering stems rise about 12 in. tall bearing 10–12 leaves arranged in two rows. The leaves are dark green tinted with red. As many as 20 or more flowers appear on each stem and are purple-red with a sweet vanilla-like scent. The flowers have prominent projections on the lower lip. In bloom June–July.

Habitat: Mostly in woodlands and scrublands of NW Yorkshire and the Lake District with its roots growing between limestone rocks. Also present in the Burren Mountains of Galway and Clare, in Ireland, growing under similar conditions.

SPIRANTHES   Tuberous-rooted perennial with a leafy stem and differing from all other British species in that its flowers are arranged in a spirally-twisted spike. It takes its name from the Greek *speira*, a coil, and *anthos*, flower. Flowers horizontal, tubular, with upper sepal and 2 petals forming upper part of the tube, the lip forming the lower part. Spur absent. Nectar secreted at base of labellum. Pollination by humble-bees and possibly by night hawkmoths which are attracted by the powerful scent and the conspicuous whiteness of the flowers.

AUTUMN LADY'S TRESSES   *Spiranthes spiralis*   Pl. 28

It forms dahlia-like tuberous roots from which arises a hairy stem only 6 in. tall. The broad, prominently ribbed, glaucous green leaves are formed in a basal rosette. They die back before the appearance of the flowering stem and, after using the food stored in the roots, they form again a short distance away from the old stem after flowering in autumn. The flowers are ivory white and form a spiral of 12–20 with downy sepals, shaded green on the inside. They have a jonquil-like scent by day and this lingers on into the night though to a lesser degree. In bloom August–October.

Habitat: Mostly confined to S England, growing in meadows along both banks of the Thames; also in Ireland in Clare, Galway and Mayo, in short grass over limestone formations.

SUMMER LADY'S TRESSES   *S. aestivalis*

A perennial with a tuberous rootstock and tubers from 1 to 4 in. long; it forms two new tubers each year. The flowering stems, about 6 in. tall, arise from a tuft of small erect leaves, pale green in colour, which supply the plant with only a small amount of chlorophyll. The small white flowers appear in a dainty slender spike of 5–20, arranged in a twisted

Plate 27
Iris Family   Yellow Flag

Plate 28
Orchid Family (i) 1 American Lady's Tresses 2 Greater Butterfly Orchid
3 Small White Orchid   4 Autumn Lady's Tresses

spiral. They are heavily scented, especially at night. In bloom July–August.

Habitat: Present only in two small areas in the New Forest of Hampshire; and in Jersey and Guernsey of the Channel Isles, in low-lying marshy ground, usually on the banks of streams.

IRISH OR AMERICAN LADY'S TRESSES
S. romanzoffiana   Pl. 28

It has thin tuberous roots and from the lateral bud formed at the base of the previous year's stem the flowering stem rises to a height of about 12 in. The narrow leaves are formed up the base of the stem, while the creamy-white flowers are borne in three spiral rows which give the spike a thickish appearance. The flowers emit a strong vanilla-like scent which is particularly pronounced at night. In bloom August.

Habitat: Present in North America from Colorado to Alaska and in Ireland from Cork and Kerry to Lough Neagh, also on the banks of the Bann in Co. Derry; in Co. Antrim and Co. Armagh; also the Hebrides but nowhere else in Europe. It is usually found in damp marshy places partially hidden by Sedges and Rushes.

NEOTTIA Saprophytic plants, the rhizomes hidden in bunches of short roots which appear above ground like the matted twigs of a bird's nest. The stems are covered with sheathing leaf-scales devoid of chlorophyll. Flowers brown, borne in a spike-like raceme, sepals and lateral petals hooded; lip 2-lobed, spur absent. Column long, slender; stigma prominent; rostellum broad, arching over stigma, exploding when touched.

BIRDSNEST ORCHID
Neottia nidus-avis

A brownish-yellow plant which feeds on decaying leaves and humus. It forms a clump of short, blunt, fleshy roots which are exposed above ground like a bird's nest and from which rises a central stem covered in leaf scales. The flowers are brownish-yellow, borne in a cylindrical spike 6 in. long. Nectar is secreted into the lower lip and can be easily reached by small flies which are attracted by the unpleasant rancid smell, like Hawthorn but without the pleasant aniseed undertones. Self-pollination is also possible. In bloom May–June.

Birdsnest Orchid

Habitat: Present throughout the British Isles in

dense woodlands of oak and beech which the sun rarely penetrates, for though the plants require moisture they require little light. Found only on calcareous and limestone soils.

GOODYERA Distinguished from *Spiranthes* by its creeping stem, from which it roots, and by its stalked leaves. Flowers borne in one-sided spiral racemes; lip pouched; spur absent; column horizontal. Pollinia formed of packets of pollen grains held together by threads. Flowers pale cream, scented. Pollination by humble-bees.

CREEPING LADY'S TRESSES *Goodyera repens*
Syn: *Satyrium repens*
Similar to *Spiranthes spiralis* but with creeping stems which increase in length each year; the 4 or 5 leaves, which are only scales until the 5th year, also increase in size as the flower stem grows longer. The erect part of the flower spike is about 4 in. tall and does not bear blooms until about the 8th year. The flowers are white and are borne in a slender twisted row, while the spike has linear bracts and glandular hairs. The flowers are sweetly scented by day and night to attract the few insects usually present in pine woods, and are visited by nocturnal insects, as well as by humble-bees during the daytime. In bloom July–August.
Habitat: In pine woods throughout central Scotland and the Orkneys but not farther south than Durham and Cumberland, though recently observed in Norfolk; possibly introduced on the roots of conifers obtained from Scotland for afforestation.

HERMINIUM Leafy perennials with ovoid tuberous roots, with one or more tubers formed at the base of the flowering stem each year, when the old tuber dies away. Leaves 2; flower spike slender; sepals and petals narrow; lip with a short lobe on each side, spur absent or very short. Column short; stigma 2-lobed.

MUSK ORCHID *Herminium monorchis* Pl. 29
It forms 2 (occasionally 3 or 4) blunt leaves and a flower spike 3–4 in. tall consisting of 10–50 small greenish-yellow flowers which have narrow sepals and petals and a very short spurred lip. The flowers emit a soft honey-like scent which attracts small bees and beetles for its pollination. In bloom June–July.
Habitat: Present across England south of the Thames, growing in the short pasture of downlands on calcareous and limestone soils. It is especially prevalent in the Cotswolds and Chilterns.

COELOGLOSSUM Perennial herbs with tuberous roots. Leaves ovate, blunt. Flowers brownish-green with long bracts, borne on a short spike; sepals and petals forming a hood. Spur short, with free nectar. Column short; stigma central; anthers joined to top of column.

FROG ORCHID   *Coeloglossum viride*
Syn: *Habenaria viridis*

It forms 2 forked tubers from which thread-like roots extend outwards. A new tuber is formed each year when the old one, together with the flowering stem, dies back. The stem grows 4–5 in. tall, with 3–4 broad leaves at its base. Each stem supports about 12 small brownish-yellow flowers, with short sepals forming a hood and a short and blunt lip spur. The long bracts often project beyond the flowers which are visited by small bees and beetles attracted by the slight honey-like scent. In bloom June–August.

Habitat: It is found at higher altitudes than any other European Orchid, at 12,000 ft in SE Europe and about 4000 ft in the British Isles. Mostly confined to N England, Scotland and Ireland, growing in short mountainous pastureland on a calcareous or limestone soil.

GYMNADENIA   Tuberous-rooted herbs with leafy stems. Flowers pink or occasionally white with spreading outer perianth segments, the inner segments forming a hood. Labellum 3-lobed; spur long, with nectar present. Column short with 2 stigmas on lateral lobes. Pollination by butterflies and moths which are alone able to reach the nectar deeply seated in what is one of the longest spurred of all our native Orchids. Flowers scented by day and night.

SCENTED ORCHID   *Gymnadenia conopsea*   Pl. 29
Syn: *Habenaria conopsea, Orchis conopsea*

It forms long slender tubers from which a stem rises 12–15 in. long, with 3–5 erect lanceolate leaves at its base. The flowers are bright magenta-pink and are produced in a dense spike. They have a powerful scent, not unlike that of Jonquils, which persists through the night. After fertilisation, the flowers take on a rancid smell. In bloom June–July.

The variety *densiflora* is a more robust plant in all respects and bears dark pink flowers which are attractively clove-scented. They are visited by butterflies. Bearing its flowers in pyramid form, it greatly resembles *Anacamptis pyramidalis* and is found in similar situation. In bloom July–August.

Habitat: Abundant on grassy slopes over limestone, throughout England and Scotland, especially in the Dovedale district of Derbyshire, in N Yorkshire and Northumberland; also in the Burren Mountains of Galway and Clare in Ireland. It is also a plant of the moorlands, found among heather.

SHORT-SPURRED SCENTED ORCHID
G. odoratissima

It resembles *G. conopsea* but has narrower leaves and a smaller flower spike. The flowers are rich pink with a spur less than half the size of that of *G. conopsea*, though the scent is equally pronounced. The lip is not as deeply lobed. In bloom July.

Habitat: Found on only one occasion (in 1912) on the Black Hall Rocks, Durham, possibly growing from seed carried there by birds from Scandinavia where it is as common as *G. conopsea* is in Britain.

LEUCORCHIS Herbs with palmate tuberous roots, like the fingers of a hand. Flowers cream, borne in dense spike; sepals and petals forming a hood. Lip 3-lobed, forming a short spur with nectar. Flowers heavily scented and, like *Gymnadenia*, visited by nocturnal Lepidoptera; also butterflies and hive-bees as the nectar is within their range.

SMALL WHITE ORCHID *Leucorchis albida* Pl. 28
Syn: *Habenaria albida, Gymnadenia albida*

It sends up a graceful, slim flowering stem 6–8 in. high which has 4–6 blunt leaves near the base. As many as 20–30 small yellow flowers tinted with green comprise the short, dense spike, while the lip is short-spurred and deeply lobed. The flowers have a powerful vanilla-like scent. In bloom June–July.

Habitat: Though present in the calcareous downlands of Sussex and Kent, it is rare in the south, becoming more abundant on the hilly pastures of Central Wales, Staffordshire and Derbyshire and northwards along the Pennine Range into Scotland. It has also been found in open coniferous forests and does not seem to have any soil preference.

PLATANTHERA Herbs with narrow tuberous roots terminating in thin horizontal fibres. Leaves broad and blunt, unspotted. Flowers borne in loose narrow spike with back sepal and two petals forming a hood over column with lip extended to form a long narrow spur. Anthers adnate to column. Flowers white, powerfully scented especially at night and pollinated only by butterflies and moths, to whose heads the pollen threads become attached.

GREATER BUTTERFLY ORCHID *Platanthera chlorantha*
Pl. 28 Syn: *Habenaria virescens, H. chlorantha*

It has 2 narrow tuberous roots with the flower stem rising to a height of 10–20 in. from the one formed the previous year. At the base are two dark green leaves about 6 in. long and stalked, while the flower spike is about 9 in. long with 12–20 flowers, larger than those of *P. bifolia* and white tinted green, with a powerful jonquil-like scent at night. The

Plate 29
Orchid Family (ii)   1 Burnt-tip Orchid   2 Pyramidal Orchid
3 Scented Orchid   4 Musk Orchid

Plate 30
Orchid Family (iii)

1 Early Purple Orchid
3 Lady Orchid
2 Red Helleborine
4 Green-winged Orchid

narrow spur is more than 1 in. long and contains nectar at the tip, which is accessible only to the long-tongued Lepidoptera. In bloom May–July.

Habitat: Present on calcareous soils in dense deciduous woodlands throughout the British Isles; rare in Scotland. Most common south of a line drawn from the Mersey to the Trent.

LESSER BUTTERFLY ORCHID *P. bifolia*

Differs from *P. chlorantha* in that its flower spike rarely exceeds 12 in. in height and is more columnar, while its creamy-white flowers are smaller and narrower. Nor is the scent so pronounced, but is more delicate and vanilla-like. In bloom June–July.

Habitat: Mostly confined to beechwoods and coniferous woods, and to places, e.g. moorlands, where the soil is of a more acid nature.

HIMANTOGLOSSUM Herbs with entire tuberous roots and 8–10 leaves which form in autumn. Lower leaves blunt and fleshy. Flowers grey-green, borne on tall leafy stems. Sepals and petals form a hood covering column. Lip long, narrow with middle lobe spirally twisted; spur short. Anther adnate to top of column. The rancid goat-like smell of the flowers attracts bees, flies and beetles with short tongues. Valeric acid is present as in Elder bark and Valerian.

LIZARD ORCHID *Himantoglossum hircinum*

Syn: *Orchis hircina*

It forms 4 basal leaves in autumn, which remain green through winter and early spring, and sends up its flowering stem to a height of nearly 2 ft with 4–6 stem leaves of deep green which become paler with age. The flower spike is almost 12 in. long with as many as 70 flowers, each of which resembles a lizard with the sepals and petals forming the head, the side lobes of the lower lip, the body, and the twisted middle lobe resembling the tail. The flowers are also of lizard colouring, greyish-green with touches of yellow, purple and red. The lip is more than 1 in. long; the spur short so that the nectar is readily available to those bees, flies and beetles with the shortest tongues. The flowers emit the rancid smell of stale perspiration rather than of goats. In bloom June–July.

Habitat: Rare, but increasing on calcareous soils in Kent and Sussex; in the Cotswolds, and NE Yorkshire at the edges of woodlands and on scrubland. Also on river-banks for it appears to enjoy moisture about its roots.

ORCHIS Herbaceous plants with two round tuberous roots unforked in British species. Basal leaves formed in a rosette and flower stem enclosed in sheathing leaves. Perianth segments (except labellum) forming a hood over column. Lip 3-lobed. Labellum prolonged backwards into a blunt hollow spur which contains no nectar though a sweet fluid is present in the tissue of the walls. Flowers mostly purple in colour and visited almost entirely by humble-bees (*Bombus terristris*) which pierce the membrane of the spur walls to suck the sweet fluid.

LADY OR WINGED ORCHID
Orchis purpurea   Pl. 30

It sends up its flowering stem to a height of about 2 ft or more from a rosette of bright green elliptical basal leaves and bears its flowers in a 6-in. spike built up in pyramid fashion. The hood is a dull purple-red, while the lip is white or palest pink often spotted with purple or brown. The spur is curved and is longer than that of other *Orchis* species. Each flower has a small brown bract. The leaves contain coumarin and as they die back in summer release the sweet smell of newly mown hay. The flowers also have a sweet perfume, resembling the almond-like scent of Heliotrope. In bloom May–June.

Habitat: Almost entirely confined to limestone soils of Kent and E Sussex, especially the North Downs where it is found on or near the edges of beechwoods or on scrubland.

BURNT-TIP OR DWARF ORCHID
O. ustulata   Pl. 29

Like a dwarf Lady Orchid, it sends up its densely packed stems to a height of only 6 in. from a tuft of 3–4 dark green lanceolate leaves. The flowers look as if they have been browned or burned by singeing, while the hood is a deep mulberry-red. The lip is white with numerous purple spots which are really tiny bunches of hairs. The spur, like that of *O. purpurea* (above), curves downwards and is blunt-ended. The flower emits the almond-like smell of Heliotrope and may be visited by butterflies, while the leaves release the smell of newly mown hay as they dry. In bloom May–June.

Habitat: Present in limestone downland pastures in SE England and in NE Yorkshire, Durham and Northumberland.

GREEN-WINGED ORCHID   *O. morio*   Pl. 30

Distinguished from Early Purple Orchid, *O. mascula*, by the lack of any markings on its leaves which remain green through winter. It is one of the most dwarf of all Orchids, rarely exceeding 4 in. in height with the spike little more than 1 in. long. The flowers are large, the sepals a deep purple-red with green veins or sometimes pink or white with the wide lip usually paler in colour. The sepals form a hood, while the long spur curves slightly upwards. The flowers are pleasantly vanilla-scented. In bloom May–June.

Habitat: Widespread on calcareous downland pastures throughout England, Wales and Central Ireland. Absent from Scotland.

EARLY PURPLE ORCHID *O. mascula* Pl. 30

The most common and widely distributed Orchid, it grows 10–12 in. tall and has several large sheathing leaves at the base of the stem which are spotted and blotched with purple. In fact, the leaves are sometimes so heavily marked that little green is noticeable. The flowers are brightest purple, borne in a loose spike, with the back sepal and petals forming a hood and with the lateral sepals turning backwards. The lip is 3-lobed and coloured yellow at the centre. The blunt-ended spur curves upwards and there are leaf-like bracts at the base of each flower. The flower is pleasantly vanilla-scented when newly opened but after fertilisation this changes to a goat- or cat-like smell with undertones of perspiration, similar to that of the Lizard Orchid, *Himantoglossum hircinum.* In bloom April–June.

Habitat: Common in deciduous woodlands and copses throughout the British Isles but growing mainly in calcareous soils and in the Home Counties and the Midlands, often accompanying Bluebells, as in the Dales of Derbyshire and Yorkshire.

History: One of the best known of all wild flowers, Shakespeare mentions it in *Hamlet,* including it with the most common of wild flowers and weeds which formed the nosegay of the crazed Ophelia:

> Therewith fantastic garlands did she make
> Of crow-flowers, nettles, daisies and long purples
> That liberal shepherds give a grosser name
> But our cold maids do Dead Men's fingers call them.

The plant was known as *Dead Men's Fingers* from the thick tuberous roots, usually two in number, with the thick flowering stem rising between them. It was also known as *Long Purples* from its elegant flowering spike. In 'The Village Minstrel', John Clare wrote of:

> Gay 'long purples' with its tufty spike
> He'd wade with shoes to reach it in the dyke.

A beverage known as *salop* was made from the tuberous roots and until the introduction of coffee into England, city *salop* houses, where the hot drink could be enjoyed, were a favourite meeting place. The early Romans knew it as *Satyrion,* believing the roots to be the food of the satyrs. Parkinson called it *Standlewort;* Coles (1657), *King's-fingers.*

The earliest of native Orchids to bloom, Mrs Charlotte Smith expressed her sorrow at the passing of the spring flowers:

> No more shall violets linger in the dell,
> Or purple orchis variegate the plain,
> Till spring again shall call forth every bell,
> And dress with humid hands her wreath again.

ACERAS Herb with entire tuberous roots and leafy stems. Flowers greenish yellow-brown, borne in narrow spike; perianth segments forming a hood over column, labellum shaped like a man with lateral lobes representing the arms, and a narrow central lobe forking into 2 segments, the legs. Column short, anther adnate to top of column. Distinguished from *Orchis* by absence of spur.

MAN ORCHID *Aceras anthropophorum*
It sends up its slender flowering stem to a height of about 12 in. and has grey-green basal leaves, which clasp the base of the stem. The flowers clothe the top half of the stem, with as many as 60 or more spaced close together, and are greenish-yellow, sometimes with the petal-edges marked with red or purple. It is one of our least attractive Orchids. The leaves are rich in coumarin, while the flowers have readily accessible nectar and emit a most unpleasant smell, like the flowers of *Cotoneaster* and *Sorbus*, to attract hover-flies and midges for their pollination. In bloom May–July.
Habitat: On calcareous downland pastures and in quarries in Kent and Sussex, Hampshire and Somerset following an imaginary line drawn from Dover to Bath. Occasionally seen in Lincolnshire in recent years.

ANACAMPTIS Herb with entire globose tuberous roots and forming 3–4 basal winter leaves. Flowers clear pink, borne in pyramidal spikes; upper sepals and petals forming a hood; lip 3-lobed; spur long and slender; nectar absent, but as in *Orchis* sweet fluid is present in tissues lining the walls of the spur. Column short; stigmas 2; anther adnate to column. Flower is unusual as by day it is sweetly scented to attract butterflies but at night the scent is replaced by the fur-like smell of several *Orchis* species to repel moths. It is thus pollinated only by day-flying Lepidoptera.

PYRAMIDAL ORCHID *Anacamptis pyramidalis* Pl. 29
Syn: *Orchis pyramidalis*
In autumn it forms 3–4 lanceolate leaves which die back as the plant comes into bloom. The flowers are borne in a pyramidal spike or cone on a stem 12 in. tall and are a lovely shade of pure shell-pink, occasionally bright red. In bloom July–August.

Habitat: Present in limestone downlands in England, Wales and S Scotland, particularly common in the Cotswolds and Polden Hills; also on the west coast of Ireland.

ARUM FAMILY Araceae

Perennial herbs, climbing or aquatic with fleshy underground stems. Flowers borne on a fleshy peduncle or spadix, enclosed in a sheathing spathe. Perianth absent or represented by scales. Stamens 1–8; ovary 1–3-chambered; fruit berry-like. The obnoxious smell given off by the brown spadix attracts midges for pollination, though the Sweet Flag, *Acorus calamus*, a semi-aquatic plant, releases a refreshing fruity scent when the stems and leaves are crushed.

ACORUS Semi-aquatic rhizomatous perennials with linear leaves and flattened scape. Spathe absent. Spadix entirely clothed with flowers; hermaphrodite; perianth segments 6; stamens 6. Ovary 2–3-celled.

SWEET FLAG *Acorus calamus* Pl. 31

A glabrous perennial resembling the Yellow Flag, *Iris pseudacorus*, in appearance, whose root yields a similar fragrant oil. However, it is distinguished from this plant by the waved margin of one or both sides of the leaf and by its spadix. The tiny flowers are yellowish-green, are packed tightly together to cover the spadix and emit a most unpleasant smell. In bloom June–July.

The flowers possess a sombre beauty, as they resemble a church spire and are greenish-brown in colour with the surface covered in a golden mosaic. They appear on stems up to 6 ft in length which, like the leaves, are fragrant when bruised.

The plant is propagated by lifting the rhizomes in November when they are divided and replanted about 8–9 in. deep at the waterside. It will not usually bloom if it is grown in gardens.

Habitat: Chiefly the Fenlands of East Anglia and Cambridgeshire growing near running water, but during recent years it has become more widespread farther inland.

History: Though the flowers have a disagreeable perfume, the leaves and stems, when crushed, release a delicious fruity scent likened by McClintock and Fitter to tangerines, by others to ripe apples with undertones of lemon and cinnamon. The powerful scent is also released when the stems and leaves are trodden upon, and they were therefore in great demand during medieval times for strewing over the floor of churches and manor houses. The *Cinnamon Iris*, as it was called, was used to cover the floors of the great cathedrals at Ely and Norwich, as the plant grew in the nearby Fenlands. Moreover, it was brought from the river estuaries of N Europe at great cost as it was always rare in Britain. One of the charges of extravagance brought against Cardinal Wolsey was that he caused the floors of his palace at Hampton Court to be strewn with scented Rushes and Flags far too frequently.

It is also found in North America, and N India where it is cultivated in the Naga Hills and where it develops a stronger fragrance in the warmer climate. It was so highly prized in earlier times that as much as £40 per acre was paid to suppliers of the London market for a single year's crop.

A volatile oil is distilled from the leaves and rhizomes, which is used in perfumery and for making aromatic vinegars, while the dried and powdered rhizomes were at one time in demand for adding to snuffs and talcum powders.

The volatile oil is contained in the outer skin of the rhizome and to peel the rhizome before it is used, as is usual in Germany, is to waste the potent part. Under microscopic investigation it can be seen that the rhizome is an open network of cells containing both oil and water passages, while certain cells contain starch granules. The oil from the root has a similar smell to that of camphor oil and it was at one time used in England to flavour and clarify beer.

When dry, the root reveals the scars of leaf-joints on the upper surface, with the scars of the roots underneath. It varies in colour from orange to dark brown, and has a spongy and cork-like interior. The smell is aromatic and pungent and the taste is bitter. In the apothecary's shop, the dry root is known as *radix calami aromatici*, and upon distillation yields 1.3%–3.9% of aromatic oil, depending upon where it has been grown. Indian root has the highest oil content.

From the root, Faust extracted the bitter principle acorin from which Thoms later obtained a neutral resin acoretin, which gives an essential oil and sugar as final products when it is reduced from alkaline solution. One can obtain a calamine solution by shaking out the acorin with ether, while the drug calamus is derived from the rootstock. In Canada and the United States the roots are sliced and boiled in maple syrup, and when dried provide a delicious candied sweetmeat.

ARUM Perennial herbs with tuberous roots. Leaves hastate, net-veined; spadix terminal; flowers monoecious, perianth absent; ovary 1-celled. Pollination is by a species of midge (*Psychoda*) which are attracted by the disagreeable smell of the flower spadix.

WILD ARUM *Arum maculatum* Pl. 31

A hairless perennial with a short, fleshy, rhizomatous root, and leaves shaped like an arrow-head, net-veined and with long stalks. They are sometimes spotted with black or brown markings and have a sheathing petiole and waved margins. In bloom April–May.

Habitat: Hedgerows and copses, chiefly south of the Thames but also as far north as Derbyshire. Less common farther north and almost entirely absent from Scotland.

History: The plant has been given more quaint names than any other, including *Cuckoo Pint; Lords and Ladies; Jack-in-the-Pulpit; Priest's Hood.*

The flowers appear early and to the countryman herald the spring-time, as the Northamptonshire poet John Clare so admirably describes:

> How sweet it us'd to be, when April first
> Unclos'd the arums leaves, and into view
> Its ear-like spindling flowers their cases burst,
> Beting'd with yellowish-white or lushy hue.

The yellow spathe is erect and is longer than the purple-brown (some-times yellowish-brown) spadix which, as the spathe opens, begins a remarkable rise in temperature. This may often be as much as 20° F above the surrounding air temperature, and at this point the spadix starts to give off a most unpleasant smell, foul and urinous which remains on the fingers and is difficult to remove. This remarkable phenomenon is a part of the flower's elaborate mechanism of pollina-tion, for the disagreeable odour is attractive only to a certain species of midge (*Psychoda*) which crawl down the spadix when the smell is at its height and past a bunch of 'hairs' (really stamens) situated at the base, immediately above the male 'flowers'. Beyond these male flowers are more hairs which turn downwards, so trapping the midges while pollination of the female 'flowers' takes place. After fertilisation, the hairs wither away to release the midges which crawl back up the spadix past the male flowers, where they collect ripe pollen to fertilise the female flowers in other Arums. The spadix begins to lose its heat and offensive smell as soon as the plant is fertilised. It quickly becomes limp and soon dies away, together with the hood, to be replaced by a bunch of brilliantly coloured berries which are extremely poisonous. The fleshy roots are also poisonous, though if they are soaked the poisonous starchy matter is removed. To exploit this process a factory was built in Weymouth (where the plant was most common) during the early years of the century for the commercial treatment and marketing of the root, which was eventually sold as *Portland Sago* and used as a sub-stitute for arrowroot.

SEDGE FAMILY Cyperaceae

Semi-aquatic perennial grass-like herbs with a creeping rhizomatous rootstock. Stems solid with sheathing leaves arranged in 3 ranks. Flowers in spikelets, borne in branched inflorescence in axil of glume.

Fruit, an achene. Flowers wind-pollinated and scentless, though the stems, leaves and roots of several species of *Cyperus* release a pleasant scent when bruised or trodden upon.

CYPERUS   Perennial rhizomatous herbs with solid, often angled stems and flowers borne in capitate inflorescence with leafy bracts; perianth absent; stamens 3, occasionally 2 or 1; stigmas 3, occasionally 2.

GALINGALE, SWEET SEDGE   *Cyperus longus*

A glabrous perennial increasing by its creeping rootstock from which a 3-angled stem rises 2–3 ft tall, clothed in recurving leaves about ½ in. wide. The reddish-coloured flowers are borne in umbel-like cymes and have long leafy bracts. In bloom July–September.

Habitat: Frequents the banks of rivers and ponds and is also found about marshy ground south of the Thames and in the Channel Islands. Very rare elsewhere.

History: From the dried rhizomatous roots, the French obtain a scented powder known as *Souchet* which is used in talcum powders, while, upon distillation, the stem and leaves yield an aromatic perfume or toilet water which has a sweet moss-like scent, resembling the Violet's though not as pure. It is a compound odour which Gerard described as 'a most sweet and pleasant smell when (the stem is) broken'. The essential oil may be mixed with lavender-water, while the roots, like orris, become more strongly scented with age and emit a similar perfume. When the outer skin of the rhizome is examined under microscope it can be seen to be composed of bundles of reddish-brown cells which contain the essential oil.

It was one of the Sedges or Rushes used during medieval and Tudor times for strewing over the floor of households of importance. A payment is recorded for 1226 of '12 pence for hay (Sweet Vernal Grass) and rushes (*Cyperus*) to strew the Baron's chamber' and in *The Taming of the Shrew*, when Petruchio sends his servant Grumio to prepare the house for his bride, Grumio calls out, 'Where's the cook? Is supper ready, the house trimmed, rushes strewed, cobwebs swept?' Later, as Parkinson tells us, Germander and Hyssop came to be planted in gardens so that the stems could be cut and used with sweet Rushes and Grasses for strewing.

Cunningham also described its value for strewing in the 'Cottage of Content':

> Green rushes were strew'd on her floor,
> Her casement sweet woodbines crept wantonly round,
> And deck'd the sod seats at her door.

Plate 31
Arum Family   1 Sweet Flag   2 Wild Arum

Plate 32
Pine Family   1 Scots Pine
Cypress Family 2 Juniper

GRASS FAMILY   Gramineae

Fibrous-rooted annual or perennial herbs varying in height from several inches to 100 ft (Bamboo), the stems circular or 2-edged, furnished with nodes. Leaves alternate, simple, arranged in 2 rows on opposite sides of stem which is embraced in sheath. The blade or upper part of the leaf is long and narrow with parallel veins. Flowers borne in spikelets arranged in spikes, racemes or panicles. Spikelets composed of alternating scales, in opposite rows one above the other, known as *upper* and *lower glumes*. The lower is called the *flowering glume*, while the inner is known as the *pale*. In certain Grasses, there is an awn attached to one or more of the glumes, often elongated as in Barley.

The flowers generally have 3 stamens with long anthers. Ovary is a single chamber with (usually) 2 styles with small feathery stigmas which collect the dust-like pollen when the stamens dehisce. The pollen is carried by wind, for many Grasses are to be found where insects are few in number. The flowers have no scent but several Grasses contain coumarin which provides the sweet herby smell to newly mown hay.

HEIROCHLOË   Perennial grasses of the cold temperate regions. Branching at each node, the spikelets are composed of 3 florets, the upper hermaphrodite with 2 stamens, and 2 lower male, with 3 stamens. Glumes membranous, 3-nerved.

HOLY GRASS   *Hierochloë borealis*

A glabrous creeping perennial of tufted habit growing 12–15 in. tall. It has a creeping rootstock, rough-edged leaves and forms a brown panicle of 3-flowered spikelets. The stems and leaves release the powerful scent of newly mown hay as they dry, and for this reason it was used in Scotland to strew over the floors of churches and castles. It takes its name from the Greek *hieros*, holy, and *chloe*, grass.

Habitat: A rare grass of N and W Scotland, inhabiting river-banks and the edges of lakes; also present around Lough Neagh in Ireland.

ANTHOXANTHUM   Annual or perennial herbs of northern temperate regions, one of which, *A. odoratum*, is rich in coumarin. Flowers in panicle of lanceolate spikelets of 3 florets, upper hermaphrodite, lower 2 sterile. Glumes thin, longer than floret, 5–7-nerved; stamens 2.

SWEET VERNAL GRASS   *Anthoxanthum odoratum*

Like *Hierochloë borealis*, a tufted perennial bearing shining stems 12–18 in. tall with flat, sparsely hairy leaves and oblong panicles of purple spikelets. The awn of the lower floret often exceeds the upper glume. In bloom May–July.

Habitat: Widespread throughout the British Isles in meadows and on moorland, also in deciduous woodlands.

History: The stem and leaves are rich in coumarin or coumaric acid, present also in Holy Grass, Melilot and Woodruff as well as in Tonka Beans. Coumarin is white, melting at 68° C. It is more soluble in warm water than in cold and crystallises into rectangular prisms. Shakespeare knew that the best hay contained Sweet Vernal Grass when he wrote, 'Good hay – sweet hay hath no fellow'.

The same satisfying herby scent is obtained by mixing the essential oil of tonka bean with that of geranium, rose and jasmine for making toilet soaps of refreshing perfume.

Perhaps this was the pleasant-smelling grass mentioned by Robert Herrick in his poem 'Candlemass Eve', for there appears to be no scent in the Ben Grasses (*Agrostis*):

> When yew is out and birch comes in,
> And many flowers beside
> Both of a fresh and fragrant kin
> To honour Whitsuntide.
> Green rushes then and sweetest bents
> With cooler open boughs,
> Come in for comely ornaments
> To re-adorn the house.

PINE FAMILY Pinaceae

Trees, rarely shrubs, with leaves spirally arranged and long shoots with scale leaves or needle-like leaves. Flowers naked, fertile flower a cone with scale-like carpels which may have in their axils other scales, or placentas, which bear the ovules. The cone becomes woody as seed ripens and remains closed until seed is fully ripe. Seeds often winged. Flowers are wind-pollinated and are unscented. An interesting note on the Monterey Pine appeared in Vol. 41 (No. 1) of the *American Horticultural Magazine*, from Mr F. W. Roe, of Ross, California. Describing recent plantings in the San Francisco Bay area he writes, 'One hot Sept. day at noon I noticed a popping, crackling noise coming from high in the pines. After careful study with the help of binoculars, I found this was coming from cones which were opening their scales and shedding seed'. In Britain, when the seed has to be extracted for sowing, this is usually done by placing the cones in a warm oven, though in a warm, dry summer the cones will shed seed in the open.

The trees are rich in pitch and turpentine which is obtained from incisions made in the bark. They give off an exhilarating smell produced by an excess of ozone and by the camphoric acid derived from the essential oil upon the decomposition of turpentine. This smell was noticed in 1875 by Mr C. T. Kingzett while he was walking in a pine wood and he was so refreshed by it that he conceived the idea of creating the same process by artificial methods for use in the home – this was the beginning of the famous Sanitas Company.

The refreshing pungent scent associated with pine-needles is due mostly to the presence of borneol acetate which is also present in Rosemary leaves.

PINUS Evergreen resinous trees of northern temperate zones, with both long and short shoots, with resinous scales on the long shoots and with needle-shaped leaves on the short ones, formed in clusters of 2, 3 or 5. Inflorescence in form of staminate cones borne in spikes; stamens 2-chambered, taking nearly 3 years before they ripen, with the seeds contained between the scales. The seeds are covered in endosperm and each has a hard, winged testa. They are dispersed by wind during warm weather.

SCOTS PINE *Pinus sylvestris* Pl. 32
An evergreen which will reach a height of 100 ft with a trunk of 4 ft in diameter, covered in warm brown scale-like bark which flakes with age. The buds are sticky and resinous, the leaves needle-like and silvery-green, usually borne in pairs. The male flowers are orange, the female solitary and pinkish, forming a cone about 2 in. long. In bloom May–June.

Habitat: On rocky and sandy ground, at first only in Scotland but now extending to all parts of the British Isles, either self-seeded or planted for afforestation.

History: It takes its name from the Celtic *pen* or *pin* meaning a rock, hence Pennine Range and Apennines where Pliny tells us grew the best pine timber, used for the foundations of Venetian houses. *P. sylvestris* will grow among rocks where the soil is so shallow that few other plants can exist. Tennyson wrote of, 'My tall dark pines, that plumed the craggy ledge'.

From earliest times, the shoots were placed in ale and the young cones were used to flavour wine. Loudon tells us that in Tuscany cones were to be seen floating in wine vats. The cones were called *pine-nuts* or *pine-apples*. Gerard knew it as the Pine-apple Tree and called the cones *Pyn Appels* for, when young, they were sweet and juicy. Shelley wrote:

> The milky pine-nuts which the autumn blast
> Shakes into the tall grass . . .

In medieval times the cones were strung up in each room of a house to impart their pungent refreshing scent and they were also placed in oak chests among clothes and vestments.

> Sweete smelling Firre that frankinsence provokes
> And Pine-appels from whence sweet juice doth come.

Parkinson tells us that the seeds 'whilst fresh and newly taken out, are used by apothecaries, comfitmakers and cooks . . . with them a cunning cook can make divers kech-choses for his master's table'.

Pitch and turpentine may be extracted from the bark, and the trees were afforded the greatest care in the past, for if they lose their bark they die back and the resin ceases to flow. Shakespeare referred to this in *Lucrece*:

> Oh me! The bark peel'd from the lofty pine,
> His leaves will wither and his sap decay;
> So must my soul, her bark being peel'd away.

The resinous roots were used by the Highlanders of Scotland for burning in their cottages as scented 'candles' for, if they were cut into thin strips, they would burn with a low flame and fumigate the house with a sweet resinous smoke.

Esters of borneol are present in all types of pine-needle oil, chiefly the acetic ester, the substance which is mainly responsible for the pine odour. The particular character of the various pine oils is also determined by the presence of different terpenes and the amount of borneol acetate present, which is usually about 5%. Present also in the oil of *P. sylvestris* is dextro-pinene and sylvestrene and, besides borneol, there may be other alcoholic constituents present.

CYPRESS FAMILY Cupressaceae

Much-branched evergreen trees or shrubs, erect or prostrate. Leaves in whorls of 3 or 4. Flowers monoecious or dioecious, borne axillary or terminal on short shoots. Stamens with short filaments. Fruit, a small berry-like cone of 1–6 scales arranged in whorls; ovules 1–2 beneath each scale. Seeds free. The Juniper is a plant of dry hilly or mountainous regions and all parts of it are pungently fragrant.

JUNIPERUS Small evergreen trees or shrubs with highly scented red wood. Leaves whorled in threes, usually glaucous above, dark green on the underside. Flowers axillary, sessile, dioecious, males discharging clouds of yellow pollen, females green, borne on scaly stalks. Fruit, a cone enveloping 3 hard seeds in fleshy endosperm, like a berry, and taking 2 years to ripen.

JUNIPER *Juniperus communis* Pl. 32
A glaucous spreading evergreen shrub or small tree bearing its awl-shaped leaves in whorls of 3. In Britain it grows 6 ft tall but reaches

40 ft in Europe. It has reddish bark which flakes. The flowers are borne at the leaf bases, male and female on separate plants, the males discharging yellow pollen which is carried to the females by the wind. The pea-like fruits are green and take 2 years to ripen, when they become deep purple with a grape-like 'bloom' and are soft and fleshy. In bloom May–June.

Habitat: Common on chalky downlands in SE England and in Scotland; *J. hibernica* is its Irish counterpart.

History: The fruits taste sweet and aromatic and have a balsamic odour when bruised. They have a high sugar content and are used to flavour the alcoholic drink known as *gin*, an abbreviation of the French *genièvre* from which the genus takes its name. Gin is a malt spirit distilled a second time with Juniper berries.

Oil of juniper consists of pinene and cadinene, but the odorous principle is considered to be the acetic ether allied to the terpenes present. The volatile oil may also be obtained from the wood and stems by steam distillation. When burnt, the yellow wood gives off a pleasing aroma and in his *Anatomy of Melancholy*, Robert Burton tells us that 'it is in great request with us at Oxford, to sweeten our chambers'. Ben Jonson also alludes to its use in his lines: 'He doth sacrifice twopence to her every morning before she rises, to sweeten the room by burning it (Juniper)'. It may also have been the 'Perfume for a lady's chamber', mentioned by Autolycus in *A Winter's Tale*, for we know that Juniper wood was burnt to fumigate the bedchamber of Queen Elizabeth. In Scotland it is used to impart its unique flavour to smoked ham.

The foliage leaves its pungent resinous aroma on the hands when it is touched and so powerful is its fragrance that the peoples of the northern hemisphere offered it as a sacrifice to their gods, while its branches were placed outside the home to prevent the entry of witches. The Hebrews burnt Juniper wood on their altars. In Exodus we find 'And Aaron shall burn thereon sweet incense every morning'. This was the wood of *Juniperus sabina* (Savin) which Ovid said was used in his time as an offering to the gods:

> The simple savin on the altars smoked,
> A laurel sprig the easy gods invoked,
> And rich was he whose votive wreath possessed,
> The lovely violet with sweet flowers dress'd.

During Evelyn's time, Juniper bushes were trained to form an arbour or bower providing a shady resting-place where one could become invigorated by its resinous smell, and spoons and forks were made from its wood, to impart their scent while eating.

Where the trees are growing in a warm climate, a gum resembling frankincense exudes from cracks in the bark and from the wood an empyreumatic oil, valuable for curing skin diseases and resembling pitch, is obtained by distillation. The flowers are also scented, which is unusual in this family. The Irish Juniper, *J. hibernica*, is a variety of upright columnar habit, with foliage of a silvery hue.

POLYPODY FAMILY Polypodiaceae

British genera are non-woody perennials (*Anogramma leptophylla*, annual) with leaves rising from a creeping rootstock. Leaves deeply lobed (only *Phyllitis scolopendrium* undivided). Reproduced from spores produced in heaps (sori) and held on the back of the leaves at the junctions of the veins. Mainly frequenting damp, shady places, several species are scented but these are usually to be found exposed to the elements, growing between rocks in full sunlight rather than in shade.

DRYOPTERIS Terrestrial Ferns of tufty habit, with lanceolate pinnate leaves rising directly from the rootstock.

RIGID BUCKLER FERN *Dryopteris villarsii*
Syn: *Aspidium rigidum*
It has thick downy blue-green leaves, 2-pinnate, the secondary leaflets with pointed teeth. It grows 18 in. tall and the leaves when handled or brushed against release a balsam-like scent.
Habitat: Mainly confined to rocky limestone ground in NW England and N Wales.

HAY-SCENTED BUCKLER FERN *D. aemula*
Syn: *Aspidium aemulum*
A handsome fern with triangular 3-pinnate leaves 18–20 in. long and a rich green, with brown leaf-stalks. It is rich in coumarin and as the leaves dry they release the refreshing scent of newly mown hay.
Habitat: Among limestone rocks and on sunny banks. Chiefly confined to the western side of England and Ireland.

Scented Plants and their Habitats

DECIDUOUS WOODLANDS

Angular Solomon's Seal	Polygonatum odoratum
Apple Mint	Mentha rotundifolia
Barberry	Berberis vulgaris
Bird Cherry	Prunus padus
Birdsnest Orchid	Neottia nidus-avis
Black Currant	Ribes nigrum
Black Poplar	Populus nigra
Bluebell	Hyacinthoides non-scripta
Box	Buxus sempervirens
Butterfly Orchid	Platanthera chlorantha
Columbine	Aquilegia vulgaris
Common St John's Wort	Hypericum perforatum
Common Valerian	Valeriana officinalis
Early Purple Orchid	Orchis mascula
Ground Ivy	Glechoma hederacea
Guelder Rose	Viburnum opulus
Herb Bennet	Geum urbanum
Herb Paris	Paris quadrifolia
Honeysuckle	Lonicera periclymenum
Lady's Slipper	Cypripedium calceolus
Lily of the Valley	Convallaria majalis
Lizard Orchid	Himantoglossum hircinum
Mezereon	Daphne mezereum
Oregon Grape	Mahonia aquifolium
Privet	Ligustrum vulgare
Ramsons	Allium ursinum
Red Dead-nettle	Lamium purpureum
Red Helleborine	Epipactis atrorubens
Roast-beef Plant	Iris foetidissima
Rowan	Sorbus aucuparia
Snowdrop	Galanthus nivalis
Spurge-laurel	Daphne laureola
Stinking Hellebore	Helleborus foetidus
Summer Lady's Tresses	Spiranthes aestivalis
Sweet Violet	Viola odorata
Turkscap Lily	Lilium martagon
Tutsan	Hypericum androsaemum
Wild Angelica	Angelica sylvestris
Wild Daffodil	Narcissus pseudo-narcissus
Wild Tulip	Tulipa sylvestris
Woodland Hawthorn	Crataegus oxyacanthoides

CONIFEROUS WOODLANDS

Creeping Lady's Tresses	Goodyera repens
Heath Bedstraw	Galium saxatile
Lesser Butterfly Orchid	Platanthera bifolia
St Olaf's Candlestick	Moneses uniflora
Scots Pine	Pinus sylvestris
Twinflower	Linnaea borealis

HEDGEROWS

Agrimony	Agrimonia eupatoria
Black Horehound	Ballota nigra
Bladder Campion	Silene vulgaris
Crab Apple	Malus sylvestris
Chervil	Chaerophyllum sativum
Crosswort	Galium cruciata
Dog Rose	Rosa canina
Dog's Mercury	Mercurialis perennis
Dogwood	Thelycrania sanguinea
Field Rose	Rosa arvensis
Fragrant Agrimony	Agrimonia odorata
Garlic Mustard	Alliaria petiolata
Hairy St John's Wort	Hypericum hirsutum
Hawthorn	Crataegus monogyna
Hedge Woundwort	Stachys sylvatica
Hoary Plantain	Plantago media
Holly	Ilex aquifolium
Hop	Humulus lupulus
Horsemint	Mentha longifolia
Lords and Ladies	Arum maculatum
Mugwort	Artemisia vulgaris
Musk Mallow	Malva moschata
Parsley	Petroselinum crispum
Pear	Pyrus communis
Peppermint	Mentha piperita
Perfoliate Honeysuckle	Lonicera caprifolium
Pink Masterwort	Astrantia major
Spearmint	Mentha spicata
Sweet Cicely	Myrrhis odorata
Sweet Violet	Viola odorata
Sweetbriar	Rosa rubiginosa
Traveller's Joy	Clematis vitalba
Wayfaring Tree	Viburnum lantana
White Campion	Silene alba
White Dead-nettle	Lamium album
Wild Strawberry	Fragaria vesca
Woodruff	Galium odoratum
Yellow Archangel	Galeobdolon luteum

MEADOWS AND CORNFIELDS

Autumn Lady's Tresses	Spiranthes spiralis
Cambridge Parsley	Selinum carvifolia
Clary	Salvia horminoides
Common Star of Bethlehem	Ornithogalum umbellatum
Corn Chamomile	Anthemis arvensis
Corn Mint	Mentha arvensis
Corn Parsley	Petroselinum segetum
Creeping Thistle	Cirsium arvense
Elecampane	Inula helenium
Field Bindweed	Convolvulus arvensis
Field Scabious	Knautia arvensis
Fritillary	Fritillaria meleagris
Golden Chervil	Chaerophyllum aureum
Meadowsweet	Filipendula ulmaria
Night-flowering Catchfly	Silene noctiflora
Ox-eye Daisy	Chrysanthemum leucanthemum
Pennyroyal	Mentha pulegium
Pepper Saxifrage	Silaum silaus
Red Clover	Trifolium pratense
Spring Crocus	Crocus vernus
Stinking Chamomile	Anthemis cotula
Sweet Vernal Grass	Anthoxanthum odoratum
White Clover	Trifolium repens
Yarrow	Achillea millefolium

DOWNLANDS, DRY BANKS AND MOUNTAINOUS SLOPES

Balm	Melissa officinalis
Bastard Balm	Melittis melissophyllum
Breckland Wild Thyme	Thymus serpyllum
Burnt-tip Orchid	Orchis ustulata
Cheddar Pink	Dianthus caesius
Chives	Allium schoenoprasum
Common Calamint	Calamintha ascendens
Corsican Mint	Mentha requienii
Creeping Rest-harrow	Ononis repens
Frog Orchid	Coeloglossum viride
Green-winged Orchid	Orchis morio
Hay-scented Buckler Fern	Dryopteris aemula
Herb Robert	Geranium robertianum
Juniper	Juniperus communis
Lady Orchid	Orchis purpurea
Lady's Bedstraw	Galium verum
Large Wild Thyme	Thymus pulegioides
Lesser Evening Primrose	Oenothera biennis
Man Orchid	Aceras anthropophorum

Marjoram	Origanum vulgare
Meadow Sage	Salvia pratensis
Moschatel	Adoxa moschatellina
Musk Orchid	Herminium monorchis
Nottingham Catchfly	Silene nutans
Purple Hawkweed	Saussurea alpina
Pyramidal Orchid	Anacamptis pyramidalis
Rigid Buckler Fern	Dryopteris villarsii
Salad Burnet	Poterium sanguisorba
Scented Orchid	Gymnadenia conopsea
Slender St John's Wort	Hypericum pulchrum
Small White Orchid	Leucorchis albida
Smooth Rupture-wort	Herniaria glabra
Snowflake	Leucojum vernum
Spignel-meu	Meum athamanticum
Spring Squill	Scilla verna
Squinancy wort	Asperula cynanchica
Western Gorse	Ulex gallii
White Horehound	Marrubium vulgare
Wild Basil	Clinopodium vulgare
Wild Liquorice	Astragalus glycyphyllos
Wild Thyme	Thymus drucei

WATER AND MARSHLAND

American Lady's Tresses	Spiranthes romanzoffiana
Angelica	Angelica archangelica
Bay Willow	Salix pentandra
Bog Myrtle	Myrica gale
Buttonweed	Cotula coronopifolia
Eau-de-cologne Mint	Mentha citrata
Fen Bedstraw	Galium uliginosum
Flowering Rush	Butomus umbellatus
Galingale	Cyperus longus
Holy Grass	Hierochloë borealis
Marsh St John's Wort	Hypericum elodes
Meadowsweet	Filipendula ulmaria
Sea Wormwood	Artemisia maritima
Stinking Goosefoot	Chenopodium vulvaria
Sweet Flag	Acorus calamus
Tubular Water Dropwort	Oenanthe fistulosa
Water Avens	Geum rivale
Water Mint	Mentha aquatica
White Water-lily	Nymphaea alba
Wild Celery	Apium graveolens
Yellow Flag	Iris pseudacorus
Yellow Water-lily	Nuphar lutea

WASTE GROUND (USUALLY IN SANDY SOIL)

Alexanders	Smyrnium olusatrum
Birthwort	Aristolochia clematitis
Borage	Borago officinalis
Breckland Catchfly	Silene otites
Caraway	Carum carvi
Chinese Mugwort	Artemisia verlotorum
Common Chamomile	Anthemis nobilis
Common Melilot	Melilotus officinalis
Common Mullein	Verbascum thapsus
Coriander	Coriandrum sativum
Dame's Violet	Hesperis matronalis
Dwarf Elder	Sambucus ebulus
Elder	Sambucus nigra
Fenugreek	Trigonella purpurascens
Feverfew	Chrysanthemum parthenium
Fragrant Evening Primrose	Oenothera odorata
Gorse	Ulex europaeus
Ground-pine	Ajuga chamaepitys
Henbane	Hyoscyamus niger
Motherwort	Leonurus cardiaca
Musk Storksbill	Erodium moschatum
Musk Thistle	Carduus nutans
Pineapple Weed	Matricaria matricarioides
Ploughman's Spikenard	Inula conyza
Scented Mayweed	Matricaria chamomilla
Slender Wart Cress	Coronopus didymus
Spiny Rest-harrow	Ononis spinosa
Sticky Groundsel	Senecio viscosus
Stinking Hawksbeard	Crepis foetida
Stinkweed	Diplotaxis muralis
Stone Parsley	Sison amomum
Tansy	Chrysanthemum vulgare
Tree Lupin	Lupinus arboreus
Wall Rocket	Diplotaxis tenuifolia
White Butterbur	Petasites albus
White Melilot	Melilotus alba
Wild Catmint	Nepeta cataria
Winter Heliotrope	Petasites fragrans
Wormwood	Artemisia absinthium

LIMESTONE CLIFF AND QUARRY

Alpine Forget-me-not	Myosotis alpestris
Burnet Rose	Rosa pimpinellifolia
Fennel	Foeniculum vulgare
Great Orme Berry	Cotoneaster integerrimus
Great Sea Stock	Matthiola sinuata

Hoary Stock	Matthiola incana
Houndstongue	Cynoglossum officinale
Pale St John's Wort	Hypericum montanum
Rose-root	Sedum rosea
Sea Holly	Eryngium maritimum
Sweet Alison	Alyssum maritimum

OLD WALLS

Buddleia	Buddleia davidii
Clove Pink	Dianthus caryophyllus
Common Pink	Dianthus plumarius
Common Snapdragon	Antirrhinum majus
Wall Germander	Teucrium chamaedrys
Wall Rocket	Diplotaxis tenuifolia
Wallflower	Cheiranthus cheiri

Glossary

Achene.	Hard, dry one-seeded fruit as in strawberry.
Adnate.	One organ united to another.
Alternate.	Where the leaves are arranged one after another; or where the stamens of a flower appear between the petals.
Annual.	A plant which completes its life cycle within 12 months.
Axil.	The upper angle formed by leaf and stem.
Axillary.	Borne at the axils of the leaves and stem.
Berry.	A fruit containing seeds embedded in its juice.
Biennial.	A plant of 2 years growth, flowering in the year following that in which the seed is sown.
Bracteoles.	Small bracts attached to the base of the pedicels.
Bracts.	Small (modified) leaves differing from the others and which are present on the flower stalks.
Calyx.	The green whorl of leaf-like organs of a flower situated below the corolla.
Campanulate.	Bell-like, as in Leucojum, Galanthus.
Capitate.	Growing in heads, as in Compositae.
Capsule.	A dry, many-seeded vessel.
Carpel.	The free or united divisions of the ovary.
Catkin.	Flowers of one sex closely crowded together, the perianths being replaced by bracts as in Willow.
Caulin.	Belonging to or produced from the stem.
Column.	A term used to denote the united stamens and pistils in orchids.
Cordate.	A heart-shaped leaf with two rounded lobes at the base, terminating to a point.
Corm.	A flat or round stem which dies back each year, leaving behind at the base a new corm developed by the action of the leaves.
Corolla.	Whorl of floral leaves known as petals, situated between the calyx and the stamens.
Corymb.	Raceme of flowers on pedicels which decrease in length as they approach the top of stem, thus bringing them on to the same level.
Corymbose.	In the form of a corymb.
Cruciform.	A flower with four petals arranged in a cross-like formation as in Wallflower.
Cyme.	Terminal inflorescence beneath which are side branches bearing a terminal flower.
Deciduous.	A tree, shrub or plant losing its leaves each autumn.
Decumbent.	Stems which lie flat on the ground but with growth arising at the tips.
Dentate.	Leaves with triangular teeth at the margins.
Dioecious.	Plants with differently sexed flowers on different plants with stamens on one, pistils on another as in Willow.
Disc.	Surface from which stamens and pistils arise; or the central florets of the Compositae as in Daisy.
Endosperm.	Nourishment for the embryo, with which it is enclosed in the seed of a plant.
Entire.	Leaves which are neither divided nor toothed at the margins.
Evergreen.	A tree, shrub or plant retaining its leaves through winter.
Exstipulate.	Without stipules.
Extrorse.	Anthers which open and shed their pollen outwards.
Filament.	The lower, stem-like part of a stamen.
Fimbriate.	Petals of flowers fringed at the margins as with Pink.
Fistular.	The hollow stems of plants as in many Umbelliferae.
Florets.	The small rayed petals of Compositae.
Follicle.	An inflated 1-celled carpel.
Genus.	A group of related species.
Glabrous.	Smooth, glossy without surface hairs.
Glandular.	A cellular secreting organ, usually raised above the surface of leaves.
Glaucous.	Leaves which have a bluish lustre, usually caused by minute hairs.
Hermaphrodite.	Where both stamens and pistil are present in a flower.
Hirsute.	Generally used to denote leaves covered in long silky hairs.
Imbricate.	Arranged over each other, like the scales of a leaf bud.
Introrse.	Anthers which open inwards towards the pistil.
Involucre.	Whorled bracts at the base of a single flower or flower head.
Keel.	The lower pair of leaves of Pea-like flowers.
Labellum.	Lip, as in Orchid.
Labiate.	Lipped; the corolla or calyx divided into two unequal parts as with the flowers of mint, thyme.
Lanceolate.	Lance-shaped leaves, tapering at each end.
Linear.	Leaves which are long and narrow.
Monoecious.	Stamens and pistils on separate flowers but on the same plant.
Nectary.	Honey-secreting organ to be found at the base of petals as in Helleborus, Fritillary.
Node.	A point on the stem where a leaf is produced and from which roots may appear in those which trail along the ground.
Nutans.	A term used to denote drooping or nodding flowers.
Opposite.	Usually a term describing leaves which grow opposite each other on the stem.
Palmate.	Segments of a leaf which spread out, like the fingers of a hand, from a central point.
Papilionate.	Shaped like the Pea flower.
Pappus.	A hairy appendage of a seed.

Parasitic. A term used to describe one plant which lives on another.

Pellucid. Transparent, usually applied to dots or glands on leaves containing essential oil.

Peltate. A term used to describe leaves when the point of attachment is on the face, not the side as is more usual.

Perennial. A plant of more than 2 years duration but usually a term for one which blooms each year indefinitely as with Primrose, Marguerite.

Perfoliate. When a leaf entirely encircles the stem to give the appearance of the stem passing through it.

Perianth. The floral parts when calyx and corolla are indistinguishable as in Tulip.

Petiole. The lower stalk of a leaf from where it joins the main stem and up to the first leaf or pair of leaves.

Placenta. Body which bears the ovules in the ovary.

Pollen. Dust or grains in the anther containing the male cells to fertilise the ovules.

Pollination. The application of pollen to the stigma.

Pome. The name given to fruits such as the Apple and Pear containing the endocarp (core) in which are the seeds.

Pratense. Growing in pasture or meadowlands.

Pubescent. Leaves and stems covered in closely and pressed hairs or down.

Raceme. Stalked flowers, borne in a spike.

Radical. Arising from just above the root as used to describe the leaves of a plant of tufted habit.

Rhizomatous. Like a rhizome or stem with a thickened base which grows horizontally usually beneath the soil.

Rosette. Leaves radiating from a central underground stem, often overlapping to form a circle.

Sagittate. Leaves shaped like an arrow.

Saprophyte. A plant obtaining its nourishment from the decaying vegetable matter in which it grows.

Segment. A term used to describe a petal of a flower or parts of a leaf divided almost to the midrib.

Serrate. Saw-edged or toothed as in certain leaves.

Sessile. Stalkless, as for certain leaves.

Simple. Used to denote those leaves which are not divided, lobed or serrate.

Sinuate. With blunt lobes, usually apertaining to leaves as with the oak leaf.

Solitary. Flowers born singly, one to a stalk.

Spadix. Succulent spike bearing numerous flowers tightly packed together as in Arum, Acorus.

Spathe. The large bract enclosing the spadix.

Spike. Like a raceme, except that the flowers are stalkless.

Spur. Extension of the lower part of a corolla as in Aquilegia.

Stamen. The male organ of a flower composed of filament and anther.

Staminode. Rudimentary organs next to the stamens.

Stellate. Star-like, radiating from the centre.

Stigma. The cellular part at the top of a carpel or style to which the pollen adheres.

Stipule. Leaf-like appendages at the base of the petioles.

Stoloniferous. A plant which reproduces itself by runners which appear at the leaf nodes.

Style. Termination of a carpel bearing the stigma.

Tomentose. Covered with silky entangled hairs.

Trefoil. A leaf composed of 3 leaflets.

Umbel. Stalked flowers arising from one point and reaching to about the same level.

Whorl. Flowers or leaves arranged in a circle of 3 or more around a stem.

Wing. The lateral petals of a Pea-like flower.

Bibliography

Title	Author	Page
Anatomy of Melancholy	Robert Burton	38, 43, 241
Book of Receipts, A	Mary Doggett	94
Book of Simples	William Bullein	184
Britannia's Pastorals, The	William Browne	14, 100, 162, 194, 213
Calendar of Gardening	Stevenson	20, 124
Catalogue of Plants	John Gerard	15, 141
Compleat Angler, The	Isaak Walton	137
Complete Gardener's Practice, The	Stephen Blake	67
Country Housewife's Garden	Wm. Lawson	67
Delights for Ladies	Sir Hugh Platt	94
Display of Heraldry, A	John Guillam	40
English Gardener, The	Leonard Meager	77
English Husbandman, The	Gervase Markham	67
Family Herbal, The	Sir John Hill	141
Flora, Ceres & Pomona	John Rea	49, 68, 138
Flora Historica	Henry Phillips	46, 153
Flora of the Midland Counties	John Purten	92
Florist's Vade Mecum, The	Rev. Samuel Gilbert	138
Flower Garden, The	John Lawrence	14, 206
Flowers of Britain	L. J. Brimble	68
Fruit Trees, Of	Ralph Austen	17, 104
Gardener's Labyrinth, The	Thomas Hill	163
Herbal	John Gerard	15, 38, 71, 94, 141
Historia Plantarum	John Rea	159
Little Herbal	Anthony Ascham	51
Mary's Meadow	Mrs Ewing	143
Names of Herbs, The	Dr Wm. Turner	184
Natural History	Pliny	19
New Book of Flowers	Maria Merian	67
New Herbal	Dr Wm. Turner	45, 66, 76, 91, 209
New Herbal	Henry Lyte	45, 65
New Orchard, The	William Lawson	192
Of Gardens, Essay	Francis Bacon	49, 52, 65
Paradisus	John Parkinson	15, 47, 49, 68, 144, 146, 163
Points of Good Husbandry, 500	Thomas Tusser	19, 51
Profitable Art of Gardening	Thomas Hyll	206
Survey of London	Stowe	76
Sylva	John Evelyn	182
Sylva Sylvarum	Francis Bacon	118
Vegetable Kingdom, The	John Hill	138

Index